ECHINODERMS

BIOLOGICAL SCIENCES

PROFESSOR H. MUNRO FOX

M.A., F.R.S.

Emeritus Professor of Zoology in the University of London

ECHINODERMS

DAVID NICHOLS

M.A., D.PHIL.

Demonstrator in the Department of Zoology
at the University of Oxford

HUTCHINSON UNIVERSITY LIBRARY

LONDON

HUTCHINSON & CO. (*Publishers*) LTD
178–202 Great Portland Street, London W.1

London Melbourne Sydney
Auckland Bombay Toronto
Johannesburg New York

★

First published 1962

This book has been set in Times New Roman type face. It has been printed in Great Britain by The Anchor Press, Ltd., in Tiptree, Essex, on Antique Wove paper.

'Non coelo tantum, sed et mari suae stellae sunt.'

J. H. LINCK, *De Stellis Marinis*, 1733
Probably the first book about echinoderms

General body plan
Reprdtve. Sys
Coelum
Systems
Physiology

CONTENTS

LIST OF FIGURES

PREFACE

While the present-day members of the phylum Echinodermata form an assemblage of animals remarkably distinct from all others, it is highly probable that in their early history they, or something very like them, provided a spring-board to chordate organization. For this reason, and also because echinoderms are such common members of both the littoral fauna and fossil-bearing rocks, the phylum is an important one for students of zoology and palaeontology. It is inevitable that in presenting a comprehensive account of a very varied group of animals within the confines of a small book some aspects are treated in greater detail than others; I am fully aware, too, that generalizations are made all too often. But my aim has been to present an introductory survey of the basic structure, adaptive radiation and evolutionary history, as far as it is known, of the living and fossil groups; to treat at greater length some problems which have a general biological interest; and to view the phylum in perspective among the rest of the animal kingdom.

All who have studied the echinoderms will know the remarkable contribution to the field by Dr L. H. Hyman in her series *The Invertebrates*. So comprehensive is her review of echinoderm anatomy, development and geographical distribution that I have treated these aspects in the briefest way only and have concentrated on other topics. Among these is the fossil history, for I believe that nowhere is there a concise summary of the evidence and interpretations of the unique fossil record of this fascinating group, and I offer this as my excuse for dwelling on the subject for several whole chapters and parts of others.

It is a pleasure to express my thanks to those friends who have so kindly helped me by reading the manuscript and suggesting corrections and improvements. In particular I thank Dr A. J. Cain and Dr J. D. Currey for the very real assistance they have given me in this way. Dr K. A. Joysey, Miss Ailsa Clark, Mr D. Heddle and Mr J. D. Woodley helped me considerably. Finally, I wish to record the assistance I have had in the preparation of this volume from the Editor, Professor H. Munro Fox, F.R.S., and from the publishers.

1

INTRODUCTION AND GENERAL FEATURES

AMONG the echinoderms we find some of the most beautiful of all sea-creatures. Who has not marvelled at the symmetrical beauty of the starfish and brittle-star, the mozaic perfection of the spherical sea-urchin, or, for those fortunate enough to have seen it, the ballet of the swimming feather-star? While the star shape, or some variant of it, is the most common body form, some members of the phylum look remarkably worm-like and others medusoid. What, then, are the features which separate the echinoderms as a phylum, distinct from superficially similar animals?

Everybody knows a starfish (Fig. 1*b*), and will probably have seen it creep, mouth downwards, rather slowly over the sea-bottom. At first sight it might appear that the five arms bend to produce movement, but a starfish can move with little or no lateral flexure of its arms. On the underside are five grooves, radiating from the mouth to the tips of the arms, and each groove contains soft appendages, the *tube-feet*. These are the organs which move the animal, by extending and bending in concert. Round the spherical sea-urchin, too, it is easy to see five columns of tube-feet (Fig. 1*e*)—in fact, it is as though the tips of the starfish's arms had been held above the centre of the animal and the arms sewn together. In the brittle-stars (Fig. 1*d*), feather-stars (Fig. 1*a*) and sea-lilies there are columns of tube-feet on each of the five arms on the same side as the mouth (*oral* *surface*), though the tube-feet, not used for locomotion here, are small and difficult to see. In the sea-cucumbers (Fig. 1*c*) five columns of tube-feet lie along the worm-like body, and those on the underside may help to move the animal.

From what has been said so far it is clear that two features of the echinoderms are outstanding: first, the adult members show a body pattern having structures present in fives (*pentamery*); and, secondly, the animals all possess tube-feet. The first feature, while nearly universal in present-day forms, has not always been so: the fossil history of the group (a history which is known as well as, if not better than, that of any other group of animals) shows us that from their first appearance there were several attempts at departing from pentamerous symmetry, none of which got very far. Today we see departures in the sea-cucumbers and in some of the urchins; but even here, where the animal has become secondarily bilateral, many of the basic structures are laid down in fives, which is rather different from the early non-pentamerous forms, in which there was no hint of basic pentamery in the arrangement of parts of the body and where some of the parts were in addition totally asymmetrical.

Surprisingly, it was not until the middle of the nineteenth century that the echinoderms were finally seen as a phylum on their own, distinct from the other principal radiate group, the Coelenterata (the polyps and medusae). The anatomists pointed out that whereas the coelenterates have a simple gut with a single hole, the echinoderms have an alimentary canal with mouth and anus, and the organs of the body lie in a distinct body cavity: the echinoderms are *coelomate*. Soon after this it became evident that there are two fairly distinct ways in which a coelom can be formed: either it can arise as a split in the mesoderm (*schizocoely*) or as an outgrowth of the gut cavity or enteron (*enterocoely*). The echinoderms belong to the latter group, and this removes them from the schizocoelous annelids, molluscs and arthropods, and relates them more closely to that group of

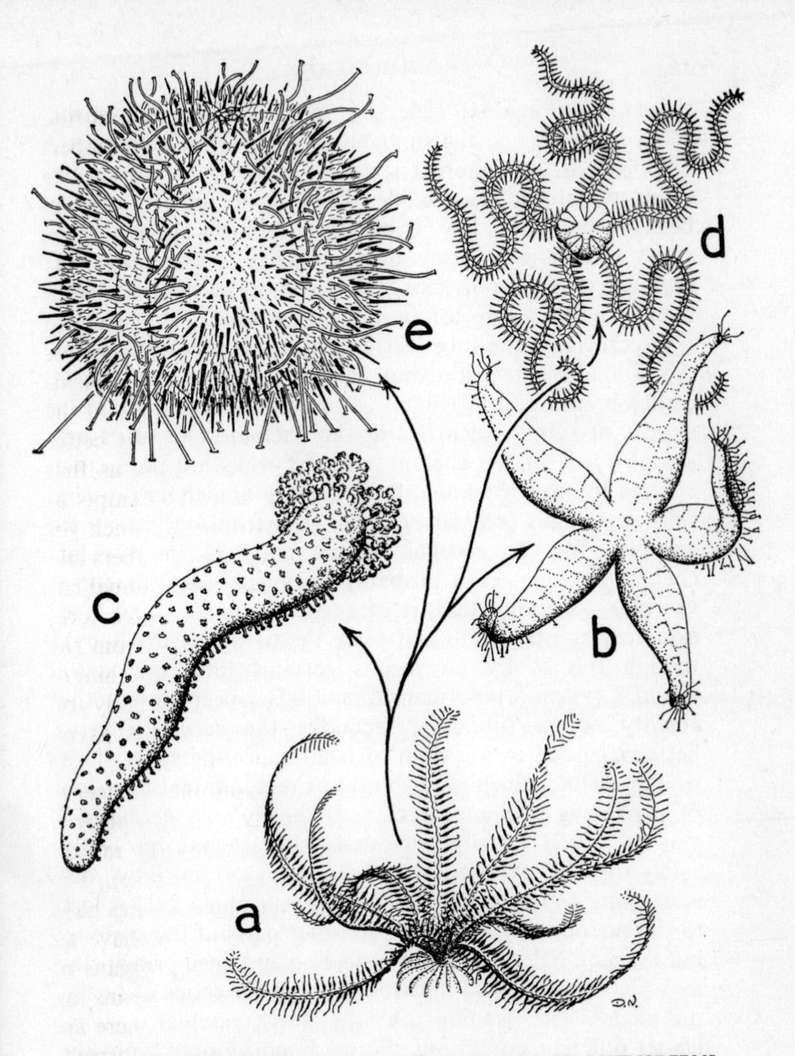

Fig. I EXAMPLES OF THE FIVE LIVING CLASSES OF ECHINODERMS

a The feather-star *Antedon*, Class Crinoidea, closest to the ancestral form.
b The starfish *Asterias*, Class Asteroidea.
c The sea-cucumber *Holothuria*, Class Holothuroidea.
d The brittle-star *Ophiothrix*, Class Ophiuroidea.
e The sea-urchin *Echinus*, Class Echinoidea.

the animal kingdom generally referred to as the *minor coelomates*, that is, the sipunculids, phoronids, ectoprocts, brachiopods, hemichordates, etc. It is still not known how fundamental the two modes of coelom formation are, because there are a few echinoderms (notably some brittle-stars) and some brachiopods which show schizocoely, but these are now generally recognized as being peculiar developmental modifications for an unknown reason.

Superficially, the tube-feet look like mere projections of the animal's outer skin, but in fact they represent blind-ending branches from a highly complex and little-understood system of tubes which lies in the coelom. The tubes are actually *part* of the coelom, and the walls are permeable enough to allow coelomic fluid and the amoeboid floating cells it contains (*coelomocytes*) to pass from one cavity to another. We can recognize three parts of the tubular coelomic system, each probably having to some extent a vascular function (that is, transportation of food and excretory substances to and from the tissues of the body), though this is by no means certain: first, the *water vascular system*, whose main function is concerned with the activity of the tube-feet; secondly, the *haemal system*, better regarded as a system of communicating spaces in a spongy tissue, which probably has the additional function of producing coelomocytes; and, thirdly, the *perihaemal system*, which usually surrounds the haemal, probably solely vascular. As the walls of the tubes are not impermeable, quite an appreciable quantity of coelomic fluid is lost to the surrounding water and to other parts of the coelom, but the required pressure seems to be maintained principally by water being sucked in through a hole or series of holes, the *madreporite*, leading into the water vascular system. Under different conditions the madreporite may probably also function as a safety-valve, as, for instance, when all the tube-feet are retracted simultaneously. The madreporite can be clearly seen on the upper surface of most starfishes and sea-urchins, and is the one external structure which breaks the symmetry of these creatures.

There is another feature in which the echinoderms differ from the other enterocoelous invertebrates, a feature which has also undoubtedly played a large part in their later success: the ability to make use of calcium carbonate from the sea-water to build a skeleton of a particularly strong kind in the mesodermal layer of the body. The earliest echinoderms in the Precambrian and early Cambrian periods of geological time built themselves a cup-shaped *test* (not called a shell, because it is an internal skeleton) with projecting food-catching processes, each having a ciliated groove on one side, and probably also a skeletal supporting rod. In this arrangement of the body both mouth and anus are directed upwards, and such a body form has persisted to the present day in the sea-lilies and feather-stars (the CRINOIDEA), but in all the other present-day forms the mouth no longer points upwards: in the starfishes (ASTEROIDEA), brittle-stars (OPHIUROIDEA) and sea-urchins (ECHINOIDEA) the mouth is directed downwards, while in the sea-cucumbers (HOLOTHUROIDEA) it is directed forwards. Such changes in attitude have inevitably meant changes in feeding habit: whereas the crinoids feed exclusively on particles of organic matter suspended in or falling through the waters around them, the others nearly all collect food from the sea-bottom. There are almost as many different ways in which they do this as there are species, but in general the primitive asteroids and most of the ophiuroids sweep material along the underside of the arms into the mouth; the holothuroids, too, mostly sweep the area in front of them with special tube-feet. But some advanced asteroids, with special suckers on their tube-feet for greater purchase, have evolved a method of preying upon molluscs to augment their diet of tiny particles. Then in the primitive echinoids we see the occurrence of teeth, used, again in conjunction with strongly suckered tube-feet, to rasp organisms from the rocks over which they move, while more advanced echinoids burrow through the substratum and take in particles of it by means of special tube-feet, then digest off the adhering organic matter as the particles pass through the gut.

No special excretory organs are found in this phylum, and the process of ridding the body of metabolic waste is something of a mystery. There is little doubt that the coelomocytes play a major role, since they have the power to move through the soft tissues, ingesting material as they go, and to discharge themselves and their contents to the exterior.

In all but a few freak forms, the sexes are separate. There is good evidence that most primitive echinoderms had a single gonad opening by a single pore close to the anus. In present-day crinoids, however, the gonads are not contained in the main part of the body, but occur in large numbers in side-branches of the arms. In the asteroids, ophiuroids and echinoids there are five gonads within the main part of the body, each opening by its own pore, while in the holothuroids there is a single gonad only. Fertilization takes place in the sea. Maturation of the germ cells is carefully timed so that a spawning animal of either sex can induce neighbouring animals to follow suit, greatly increasing the chances of fertilization. In addition, it seems that many echinoderms are gregarious, to their obvious advantage in reproduction. Development in by far the greater majority is indirect; that is, they pass through a larval stage. Since the adults are somewhat sluggish, the larvae are the main dispersive phase of the animal and remain in the plankton for sufficient time to be swept from the place of their birth to new areas, or to restock the original areas.

Apart from being a phylum with a world-wide distribution and with considerable diversity of structure on the basic plan, the echinoderms occupy a unique place in zoology: it is generally agreed nowadays that in them, or at least very close to them, we have the ancestors of the great phylum of the Chordata, at the head of which man himself sits. The way in which these two superficially quite dissimilar phyla are related is one of the most exciting speculations in the whole field of zoology, and the manner of their connexion is dealt with at the end of the book.

It now remains to consider the groups which make up

the phylum Echinodermata. The most useful first subdivision is into two *subphyla*, the Pelmatozoa (Greek: 'stalked animals') most of which are permanently attached to the sea-bed when adult, and the Eleutherozoa (Greek: 'free animals') which are free-living. Present-day forms are divided into five easily recognized *classes* within the phylum:

1. *Crinoidea*, the sea-lilies and feather-stars, the most primitive living class and the only pelmatozoan class with living members. Greek: 'lily-like'.
2. *Asteroidea*, the starfishes (or sea-stars) and cushion-stars, the most primitive eleutherozoan class with living members. Greek: 'star-like'.
3. *Ophiuroidea*, the brittle-stars and basket-stars. Greek: 'snake-like'. These and the asteroids are sometimes combined as the class Stelleroidea, when they themselves become sub-classes.
4. *Echinoidea*, the sea-urchins, sand-dollars and heart-urchins. Greek: 'spiny'.
5. *Holothuroidea*, the sea-cucumbers. Greek: 'sea-cucumber'.

In addition, past seas have seen representatives of the phylum which cannot usefully be fitted into the framework of the above classes, so the following totally extinct classes are usually recognized:

6. *Cystoidea*, sessile cyst-like forms, regarded by some as the simplest (but not necessarily the most primitive) echinoderms. Greek: 'bladder-like'.
7. *Blastoidea*, sessile bud-like forms. Greek: 'bud-like'.
8. *Edrioasteroidea*, mostly free forms which are considered to be of importance in the ancestry of the recent Eleutherozoa. Greek: 'sessile starfish'.
9. *Heterostelea*, asymmetrical forms usually with a stem or anchor. Greek: 'different stem', referring to the fact that the stem is composed of several columns of plates rather than just one column, as is more usual.

10. *Ophiocistioidea*, free forms with gigantic tube-feet on the underside. Greek: 'snake capsule'.

Finally, there are three groups, each of which has been raised to class rank by some workers, but which are only reluctantly included in the phylum at all, for reasons explained in Chapter 12:

11. *Machaeridia*, bilaterally symmetrical worm-like remains with a skeleton of imbricating plates. Greek: 'sabre-like'.
12. *Cycloidea*, radially symmetrical, non-pentamerous forms. Greek: 'rounded'.
13. *Cyamoidea*, biradially symmetrical forms otherwise very similar to the cycloids. Greek: 'bean-like'.

It is fairly certain that the original echinoderm was very different from those which move freely, though somewhat slowly, over the bottoms of modern seas. It remained attached to the sea-bottom while feeding on the rain of detritus falling to it from the waters above. It had a cup-shaped protective test, and food-collecting arms which may have been supported by a skeleton. The living class whose members most closely resemble this archetype is the Crinoidea, and among this group are found those echinoderms which appear first in the fossil record. So we shall consider this class first in the survey of the living forms which follows. Then, some of the important general features of the living forms will be dealt with, followed by a brief survey of the fossil groups. Finally, the phylogeny of the echinoderms and their relationships in the animal kingdom will be considered.

2

THE CRINOIDEA

CONVENIENTLY enough, the crinoids (sea-lilies and feather-stars), in addition to being the first echinoderms to appear in the fossil record, have retained a primitive structure throughout their evolutionary history, and there are living representatives which are quite easily obtained for study. So they make the obvious starting-point in our brief survey of the classes. The fossil record tells us that the stemmed crinoids flourished in the Palaeozoic, particularly in the Carboniferous period, some rocks of that age consisting of little more than the fossil remains of crinoids; since then the stemmed forms declined in importance and were superseded by crinoids that become secondarily free during their life-history, the comatulids or feather-stars; these are today far more widespread than their less mobile forbears, particularly in coastal waters. The best-known example, *Antedon*, is a comatulid, though its basic anatomy is very similar to that of stalked forms.

General body plan

The typical crinoid body is made up of a *stem* with some sort of holdfast, a cup-like calyx or *theca*, housing the lower part of the body, a domed flexible roof, the *tegmen*, housing the rest of the body, with the mouth at its centre, and a series of *arms* arising from the sides of the theca; each arm usually bears a large number of *pinnules*, branching alternately from its sides.

The stem, seldom exceeding two feet in modern forms but sometimes reaching seventy feet in the past, consists of a vertical series of round or star-shaped ossicles embedded in epithelium, and bears branches or *cirri* which are retained on the very reduced stem of comatulids for temporary fixation. At the top of the stem is a single plate, the *centro-dorsal*, forming the base of the theca. This is the only stem ossicle to be retained in the comatulids. The sides of the theca are made up of two whorls of five plates, the *basals* aborally and the *radials* orally. In some forms an additional whorl, the *infrabasals*, is intercalated between the basals and the centro-dorsal. In *Antedon* the thecal plates are very reduced and modified to form a capsule enclosing the chambered organ (part of the coelom, see p. 28) and the most important part of the nervous system; the five basals are reduced, fused together and form a *rosette* which acts as a cover to this capsule (Fig. 2). There is fossil evidence that primitively the tegmen contained a single whorl of five plates, and in some primitive living forms, such as *Hyocrinus*, these are still visible close to the mouth, though extra plates are intercalated between them and the radials. In less primitive forms the five plates disappear and small extra plates are scattered over the whole tegmen.

The arms arise from the boundary between the calyx and the tegmen—actually, they are borne on the radial plates of the calyx. There are five arms primitively, and this number is retained in a few forms, such as the living *Ptilocrinus*; but more often each arm divides into two, and sometimes this forking is continued again and again, the number of arms reaching about sixty in some stalked forms, e.g. *Metacrinus*, and even up to 200 in some comatulids, e.g. *Comanthina*. The number of arms appears to be determined by food conditions, and Clark[14], who has monographed the existing crinoids, makes the interesting point that the number of arms in comatulids seems to be correlated with the depth and temperature of the seas in which they live, those in shallow warm water tending to have around forty arms, and those in deep, colder water

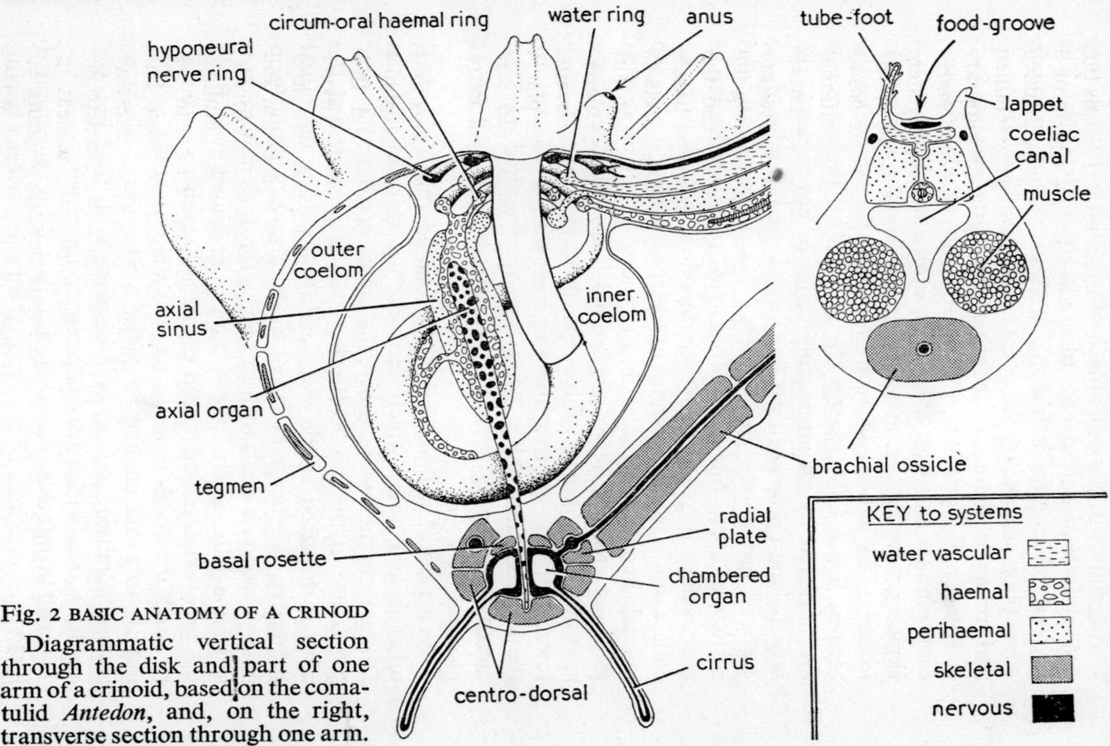

Fig. 2 BASIC ANATOMY OF A CRINOID

Diagrammatic vertical section through the disk and part of one arm of a crinoid, based on the comatulid *Antedon*, and, on the right, transverse section through one arm.

having only ten; those having twenty to thirty arms tend to occupy intermediate depths of moderate temperature. The skeletal pieces supporting the arms are called *brachials*, while those supporting the pinnules are *pinnulars*. The arms and pinnules represent the food-collecting system, outclassing anything seen elsewhere in the living echinoderms: each pinnule has a groove on its oral side leading into a similar groove in the arm which bears it, and the arm groove in turn leads into a system of grooves in the surface of the tegmen which converge on the mouth. All these grooves are bordered by an alternating series of cover-plates, the *lappets*, and groups of tube-feet arise from the sides of the grooves just inside the lappets. The arrangement and activity of these tube-feet is discussed fully in Chapter 9; here, suffice it to say that their function appears to be to collect food from the rain of detritus falling upon the animal, bundle it up in strings of mucus and convey it to the food grooves. When the animal is disturbed by some creature that might nibble the delicate tube-feet, they retract into the groove and are immediately covered by the lappets. Presumably, while they are retracted the food already in the groove system can continue its journey to the mouth.

As is general in almost all sessile echinoderms, and indeed other invertebrates which feed on the rain of detritus falling from the waters above, the mouth is situated as near the centre of the food-collecting area as possible and the anus is displaced to the side, slightly upsetting the radial symmetry. Discharge of faecal waste, always a problem in ciliary feeders, is dealt with by having the anus mounted on a flexible spire, so that the waste can be aimed away from the grooves; some advanced comatulids even go so far as to send the food grooves on a circuitous route to the mouth, to avoid passing too close to the anus; others, as though admitting defeat, even dispense with food grooves on those arms which arise adjacent to the anus.

In all living echinoderms the mouth is always surrounded by five (or multiples of five) special tube-feet usually having a mainly sensory function. In crinoids, as in most of the others, these tube-feet arise from a circum-oral water vascular ring vessel. Their turgor is maintained by the pressure in the water vascular system, and this, in conjunction with longitudinal muscles, enables them to bend to taste food passing towards the mouth. The mouth leads into a short oesophagus and this into an intestine which either has a single loop with many diverticula, or makes four turns and lacks diverticula. There is a short rectum, whose walls are muscular and which can pulsate, probably causing water to be taken into the rectum, either as an enema or for anal respiration, a process which certainly occurs in some other echinoderms (p. 80).

Reproductive organs

We saw in Chapter I that in early echinoderms there was almost invariably a single gonopore opening on the side of the theca, inferring that there was a single gonad only, and this was housed in the theca. But in the crinoids the reproductive system is entirely different: there are many gonads, situated on the proximal pinnules of each arm (except the first pair, which usually act to protect the tegmen) and extending to well over half the pinnules of the arm. Only by close examination are the sexes distinguishable. The gametes are shed into the sea by rupture of the pinnule wall and usually development proceeds in the water, or sometimes while the egg is attached to the parent by a sticky secretion, unless a brood pouch is present on the arms (e.g. *Notocrinus*) or on the pinnules (e.g. *Isometra*).

Excretory system

It is most likely that the main method of excretion, as in other echinoderms, is by coelomocytes. In crinoids the full coelomocytes collect in special spherical *sacculi*, which lie in rows alongside the ambulacral grooves, and it has been said that the sacculi discharge their waste into the sea at intervals. An interesting thing about these organs is that they also seem to collect pigment granules from the body tissues when the animal dies, so that in preserved specimens they appear darkly coloured.

The coelom

In all echinoderms the coelom is subdivided very early in development (p. 120) into parts destined to become, on the one hand, the perivisceral coelom (that surrounding the main organs of the body) and, on the other, the tubular systems. The main subdivision of the perivisceral coelom is by a membrane parallel to the wall of the theca, dividing the cavity into *inner* and *outer* portions; of these the inner is restricted to the theca, while the outer continues into the arms and pinnules as the *coeliac canals* (Fig. 2). In the walls of these canals are tiny *ciliated pits*, which may be the places where coelomocytes pass into the surrounding tissue.

The tubular elements of the coelom are as follows:

1. *The water vascular system.* Although of a similar plan to that of other echinoderms, this system in crinoids has no direct connexion with the exterior, that is, there is no madreporite as such. Instead, water passes first into the perivisceral coelom, through *ciliated funnels* in the tegmen, and from there it passes through a large number (150 in *Antedon*) of tiny ciliated water pores in the interradii of the circum-oesophageal water ring. This ring gives off a pair of oral tube-feet in each interradius (mentioned previously) and a radial water vessel in each radius, which divides where necessary to pass into each arm; then branches go to each pinnule, and the canals in arms and pinnules give off branches to the tube-feet, where tube-feet occur. It seems to be fairly general among the crinoids that the water vascular canals, including the circum-oral ring, are traversed by muscle fibres, the function of which is dealt with in Chapter 9.

2. *The haemal system.* This is a somewhat indefinite system

of intercommunicating spaces, rather than a system of tubes, sending elements to all parts of the body. It consists of a number of rings surrounding the first part of the gut, an axial portion, and a branch to each arm which sends 'twigs' to each pinnule and tiny channels to each tube-foot. Round the first part of the oesophagus is the *circum-oral haemal ring* which is in communication by many channels with the *genital haemal ring* just below it, from which the *radial haemal strands* arise; these, passing along each arm just above the coeliac canal, give off branches to the pinnules, and it is these branches, greatly expanded, which contain the gonads (hence the name of the ring round the gut from which they arise). Also arising from the genital haemal ring is the so-called *spongy body*, an extensive system the main part of which is in close association with the gut walls. The spongy body is thickest near the 'hub' of the animal, that is, at the axial complex, where it surrounds an *axial organ*. We can conveniently consider this organ as part of the haemal system throughout the phylum, because it is always in close association with it, the channels of the haemal system ramifying within it. Aborally, the axial organ continues further than the spongy body (Fig. 2), disappearing, in fact, only after it has passed through the rosette formed by the five fused basal plates. During development the axial organ sends a process along the middle of each radial haemal strand to the pinnules, processes which later enlarge to form the gonads. The connexion between organ and processes is broken later in life, but many early workers thought that the germ cells originated in the organ, and hence called it the genital stolon. Other workers reported seeing it pulsate (as do most of the tubular coelomic parts, anyway) and because of this called it the heart.

3. *Perihaemal system*. As its name suggests, this system normally surrounds the haemal complex, though some recent authors prefer to call it the *hyponeural sinus system*, referring to its relation to one part of the nervous system. This system, like the preceding two, has a circum-oral ring, the *perihaemal ring*, an axial part, the *axial sinus*, and a radial

branch to each arm and thence to each pinnule, the *subtentacular canals*. In each arm the canal is subdivided by a vertical septum, in which runs the haemal strand (Fig. 2, transverse section).

In addition to these three tubular coelomic systems, present in all echinoderms, there is in crinoids another coelomic structure aborally, called the *chambered organ*. This occupies the region of the body enclosed by the thecal plates (Fig. 2) and sends branches to the cirri.

So far, except for the water vascular system, no function has been mentioned for the tubular coelomic structures. Indeed, this is one of the outstanding problems of echinoderm physiology; for we do not know with any certainty the functions and functional relations of any of them. One thing we can say: every major part of the nervous system has a coelomic tube or tubes close to it (e.g. in crinoids the radial nerve cords have the radial perihaemal vessels, and the aboral nervous system has the chambered organ); and branches of some tube systems are intimately related to the gut. So we can say with some conviction that all are concerned in some degree with transport of essential materials. And we know that all are richly supplied with coelomocytes, which probably carry excretory products.

The nervous system

One of the features of the echinoderm nervous system is that a mainly sensory nerve plexus lies just beneath the external epithelium of almost the entire body. This may enlarge locally where special sensitivity is required, as, for instance, at the tips of some tube-feet and on the cirri. In addition to this general nerve net there is an extensive 'central' system also. In crinoids this is in three main parts, each intercommunicating. The main sensory part lies orally, and has been called the *superficial oral* or *ectoneural* system; it is little more than an expansion of the subepithelial plexus in a ring round the mouth and beneath the centre of the ambulacral groove. Just below (aboral to) this is another sensory system, the *deeper oral* or *hyponeural* system, also consisting of a ring round the gut and branches in each arm; this time there are two to each arm and pinnule, and they lie laterally. Then the main motor system in crinoids, concerned with posture and movement[17], is the *aboral* system, which surrounds the chambered organ within the thecal plates and has a ring embedded in the radial plates; the ring gives off a brachial nerve to each arm and pinnule, situated in a canal within the brachial or pinnular ossicles. Lateral connexions run to the lateral brachial nerves of the deeper oral system.

We may summarize the basic anatomy of crinoids as follows: they have a stem by which they are attached to the sea-floor for at least part of their life-history; their food is collected by an extensive system of arms and passed by cilia to an upwardly directed mouth. Circulation of essential materials and excretory products is probably carried out by the tubular coelomic systems which ramify through the animal.

The Eocrinoidea

Before dealing with the evolution of the crinoids proper, mention must be made of a much-neglected but phylogenetically important group of early echinoderms, the Eocrinoidea, which are regarded by some [18, 95] as the basal stock from which all subsequent echinoderms arose. Actually, they are probably sufficiently distinct morphologically to be regarded as a separate class in their own right, but here they are placed as a sub-class of the Crinoidea, as are their near-relatives, the Paracrinoidea. Their stratigraphical range, Lower Cambrian to Middle Ordovician, and many of their structural features are certainly consistent with the view that they gave rise to the Cystoidea (Ordovician to Devonian), Paracrinoidea (Middle Ordovician) and Heterostelea (Middle Cambrian to Lower Devonian) as well as to the true Crinoidea which do not appear until the Lower Ordovician.

The first eocrinoid to appear is *Eocystites* from the Lower Cambrian of North America and Europe, and this may well be the earliest echinoderm known from the fossil record. Unfortunately but understandably, this form is poorly known: it is said to have ten brachioles around its upwardly directed mouth and a theca of irregular plates. Far better known is the Middle Cambrian to Ordovician *Macrocystella* (Fig. 3*a*) in which the theca is composed of four whorls of large plates with five brachioles, each of which immediately divides into two. Each branch consists of two parallel columns of plates and bears a food groove on its oral surface; the grooves are continued on to the oral face of the theca and lead to the mouth at the centre. This genus has a long stem, but another, *Lichenoides* (Middle Cambrian), appears to lack a stem altogether.

As regards respiratory structures, there do not appear to be any perforations through the thecal plates. Probably there was a canal on the oral side of each brachiole, across the walls of which gaseous exchange took place, though no trace of this system remains.

Fig. 3 EVOLUTION AND ADAPTIVE RADIATION IN THE CRINOIDS

a The eocrinoid *Macrocystella* (L. Camb.), one of the earliest of all echinoderms.

b to *d*, Inadunata.
b *Dendrocrinus* (L. Ord.), the oldest true crinoid.
c *Petalocrinus* (Sil.), with arms expanded into food-collecting flanges. Arm nearest reader removed.
d *Hybocystis* (Ord.), in which two of the ambulacra pass down the theca instead of being borne on arms.

e and *f*, Camerata.
e *Platycrinites* (M. Sil.—M. Perm.), with irregularly branching arms.
f *Barrandeocrinus* (M. Sil.), in which the pinnules bend to form food-collecting troughs. The two arms nearest the reader removed.

g and *h*, Recent Articulata.
g *Ptilocrinus*, a stalked form.
h *Antedon*, a comatulid.

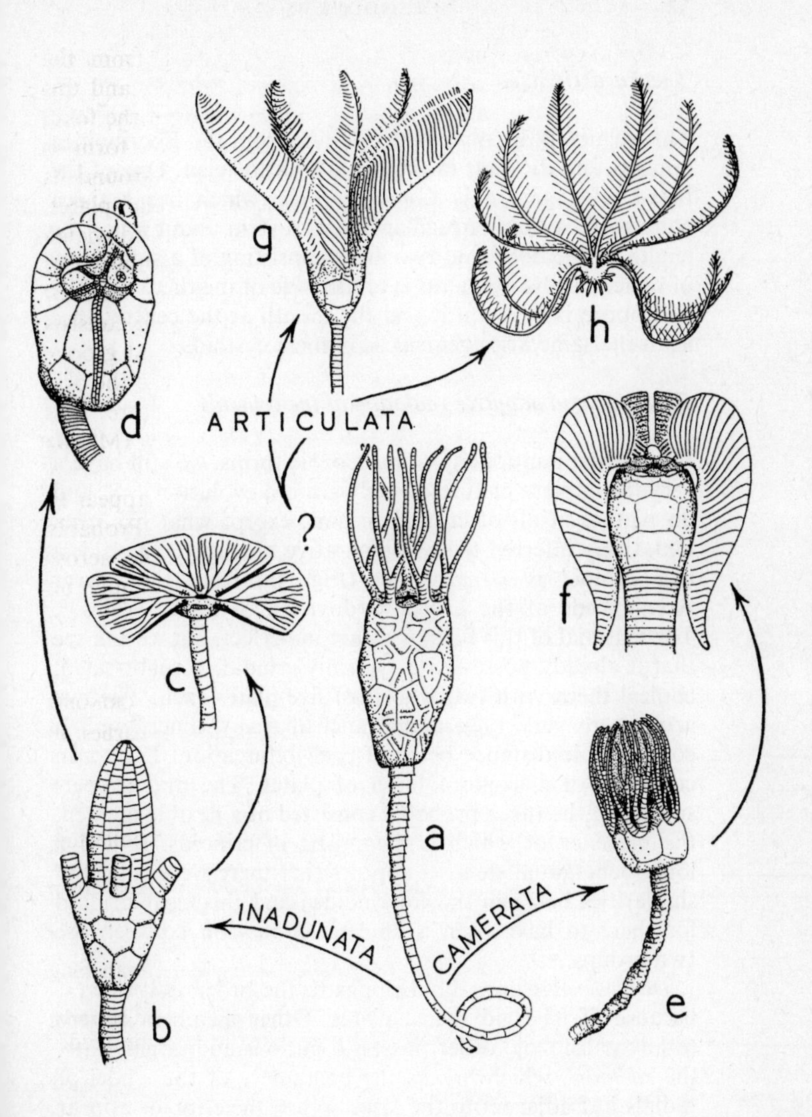
g
h
d
ARTICULATA
?
f
c
a
b
INADUNATA
CAMERATA
e

The Paracrinoidea

This relatively unimportant side-line from the Eocrinoidea appears and dies out in the Middle Ordovician. One of the best-known genera is *Comarocystites*, which has a theca composed of many irregular plates, a stem about twice the length of the body and two arms consisting of a single row of ossicles only. The anus is on the side of the theca and the hydropore is between it and the mouth at the centre; that is, in the same arrangement as in most cystoids.

Evolution and adaptive radiation of the crinoids

Despite the multiplicity of palaeozoic forms, we still have a very incomplete picture of early crinoid evolution [18, 19], and no picture at all of crinoid origins, except what has been tentatively inferred from comparative anatomy. The oldest true crinoid is *Dendrocrinus* (Fig. 3*b*) from the Lower Arenig beds of the Lower Ordovician of South Wales [10]. The material of this form is rather imperfect, but we can see that it already possessed a typically crinoid, straight-sided, conical theca with two whorls of five plates each. The five arms were very long and branched many times, with a considerable distance between each bifurcation. Each arm consisted of a single column of plates. The oral (upper) surface of the theca probably consisted of a flexible tegmen, the presence of which is diagnostic of crinoids. It had a long, pentagonal stem. It appears that there were sufficient similarities between the Eocrinoidea and this early crinoid for there to have been a phyletic connexion between the two groups.

Dendrocrinus probably belongs to the order INADUNATA, because of its rigid thecal plates. Other members of this totally palaeozoic order possess a curious additional plate, the *radianal*, which breaks the pentamery of the whorl of radials just adjacent to the arms. When the crinoids appear in quantity, a little later in the Ordovician, two other orders can be recognized: the FLEXIBILIA, with flexible theca, and the CAMERATA, with rigid theca but no radianal plate and with some of the proximal arm ossicles incorporated into the theca.

Though these three palaeozoic orders are generally agreed to have been distinct, certain evolutionary trends are common to all of them. For instance, the primitive crinoid theca is conical in shape, with straight sides, as in *Dendrocrinus*: during evolution the base of the cup becomes flat, then concave, then finally surrounds the top of the stem. The ratio of height to width decreases.

Some authorities[1] consider the number of whorls of thecal plates below the ring of radials, either one (monocyclic) or two (dicyclic), to be important taxonomically, but most other workers consider that this arrangement is misleading, because this character appears again and again in the group, and in some genera, e.g. *Uintacrinus*, some species are monocyclic while others are dicyclic[4], so that Bather's[1] sub-classes Monocyclica and Dicyclica can profitably be rejected.

Among early inadunate crinoids are forms such as *Hybocystis*, Ordovician (Fig. 3*d*), in which only three of the five ambulacra are continued on to arms: the food grooves of the other two ambulacra continue down the sides of the theca. In other forms, e.g. *Petalocrinus*, Silurian (Fig. 3*c*), the arms are expanded to form fan-like food-gathering structures, over the oral surface of which the food grooves ramify extensively.

There is less radiation among the flexible crinoids[20], most forms having the typical lily-like appearance. To mention just two, the earliest flexibles are *Protaxocrinus* (Middle Ordovician) with arms that branch dichotomously and *Ichthyocrinus* (Silurian) with many plates in each whorl and each arm bifurcating. Of the camerates, a typical form is *Platycrinites*, Middle Silurian to Middle Permian (Fig. 3*e*), with irregularly branching arms and a spiral stem. In many camerates it appears that the arms sagged down over the theca, like a wilting lily. This tendency is taken to an extreme in *Barrandeocrinus*, Middle Silurian[9] (Fig. 3*f*), in which the arms hug the sides of the theca and the pinnules of each arm bend towards each other to form a trough into which, apparently, water and food particles were admitted between the pinnules, then taken to the mouth.

None of the three foregoing orders passes into the Mesozoic. In the Triassic the fourth and last order, ARTICULATA, appears, in the members of which the tegmen is flexible and the basal ossicles of the arms articulate with the radial plates of the theca. In the Lower and Middle Triassic all articulates have stems (an example is *Pentacrinites*), but stemless forms, the Comatulida, first appear in the Upper Triassic of Mexico. This group, in which advantage is taken of the arm articulation for locomotion, appears to be polyphyletic; that is, several groups within the Articulata became free independently. In the Jurassic the comatulids are still rare, but in the Cretaceous they flourish, and they appear to hold their superiority, at least in littoral waters, until the present day.

There has been much discussion about the systematic position of one recent form, *Hyocrinus* from Antarctic waters, which has the straight-sided conical theca reminiscent of very early crinoids of the order Inadunata. At least one textbook[4] links it, via a couple of Jurassic forms of similar appearance, to the otherwise totally palaeozoic Inadunata, but recently it has been suggested[18] that its primitive features have arisen by convergence, and that it really belongs, like all other living crinoids, to the order Articulata.

3

THE ASTEROIDEA

THE starfishes, the first of the eleutherozoan classes we shall consider, are famous objects of the seashore and infamous visitors to oyster-beds, where their ability to wrap themselves round the bivalves, tug at the shell against the muscles which close it and extrude their stomachs to digest away the meat has exasperated fishermen and fascinated zoologists for years. In the early days of oyster-farming the fishermen would drag a dredge across the beds to collect the starfish, tear them in pieces and throw them back again; they did not know then that there was a very good chance of every piece regenerating lost parts, considerably increasing the menace in time.

During the early part of their life (p. 125) the asteroids become fixed to the sea-bottom for a short time, and this has been considered good evidence for regarding them as fairly close to the Pelmatozoa, and probably among the first Eleutherozoa to break free and invert; there are also features in their adult anatomy which indicate their rather primitive position (p. 168).

Though by far the majority of starfishes are pentamerous, the class exhibits greater departure from this number of rays than any other. For instance, the Atlantic species *Luidia sarsi* has the normal five arms, while *L. ciliaris* has seven; the Sunstar, *Solaster*, may have any number from fifteen to fifty.

General body plan

Basically, the asteroid body consists of a central disk with the mouth in the middle of the undersurface and anus in the centre of the upper. A number of arms project laterally. The delimitation of arms and disk is not very marked; indeed, in some, such as the Duck's Foot, *Anseropoda* (Fig. 5*j*), the outline is roughly that of a pentagon. The main part of the alimentary canal is situated in the disk (Fig. 4), but pouches extend into each arm. Similarly, the central parts of all the tubular coelomic systems are found in the disk, with radial extensions along each arm. All asteroids have a skeleton of internal plates, but while in some they may be closely packed and form a complete layer in the body, in others some of the ossicles, particularly those of the upper surface, have large spaces between them. Round the mouth is a special *peristomial ring* of closely abutting plates, four *ambulacrals* in each radius alternating with two *adambulacrals* in each interradius. Oral to each pair of adambulacrals there is usually a single plate, called the *interradial*. The elements of this ring represent the first of the series which extend out in the oral surface of the arms, each arm having two parallel columns of closely packed ambulacrals with a single column of adambulacrals on each side of them. This explains why the adambulacrals cannot be called interambulacrals—each pair round the mouth is split between two adjacent radii. The ambulacral plates normally form a deep groove in the oral surface of each arm and the plates have pores between them, often in a zig-zag, taking canals from the tube-feet to ampullae

---

Fig. 4 BASIC ANATOMY OF AN ASTEROID

Diagrammatic vertical section through the disk and part of one arm of a starfish, based on *Asterias*. On the right, transverse section through one arm, showing one tube-foot and its ampulla. A section of the axial complex is shown at lower left.

See Fig. 2 for key to systems.

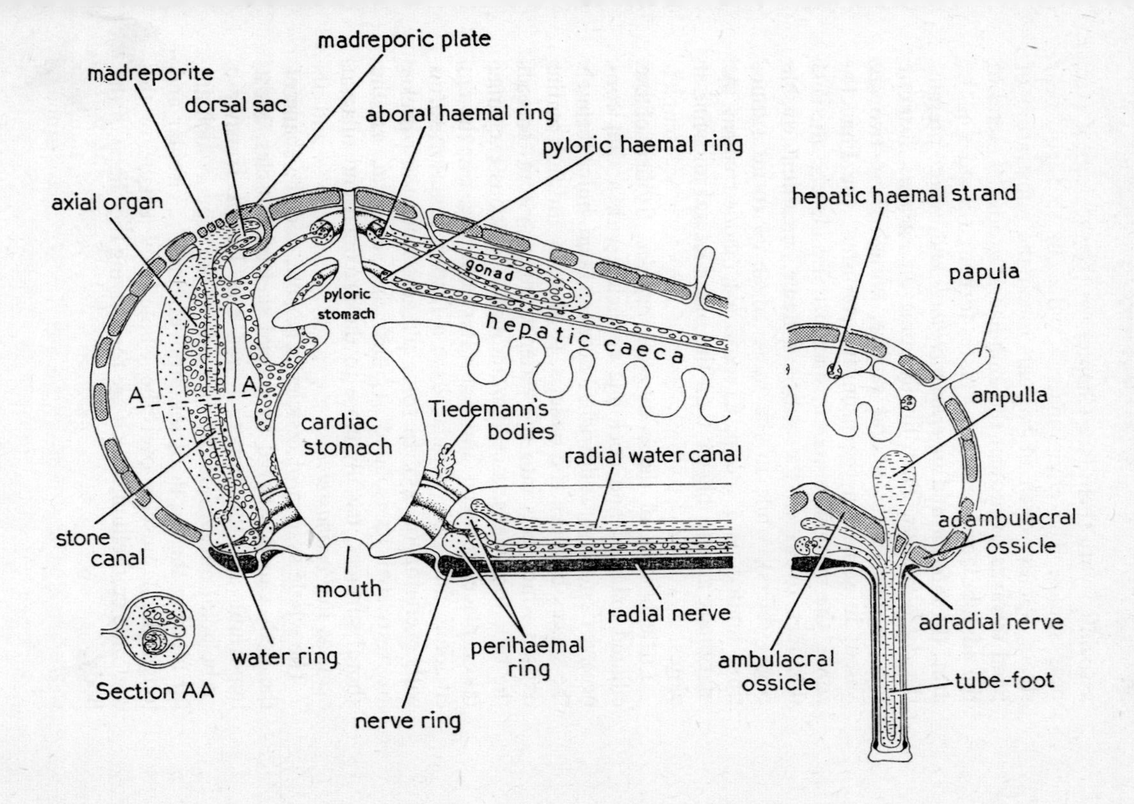
madreporic plate
madreporite
dorsal sac
aboral haemal ring
pyloric haemal ring
axial organ
hepatic haemal strand
gonad
papula
pyloric
stomach
hepatic caeca
ampulla
A
A
cardiac
stomach
Tiedemann's
bodies
radial water canal
adambulacral
ossicle
stone
canal
mouth
radial nerve
adradial nerve
perihaemal
ring
water ring
ambulacral
ossicle
tube-foot
Section AA
nerve ring

associated with them (see Chapter 9); they seldom, if ever, bear spines. The adambulacrals, adjacent to them, bear strong spines which in life can touch the substratum or bend inwards to protect the ambulacral groove. Lateral to the adambulacrals come the plates forming the sides of the arm: the *inferior* and *superior marginals* and, where present, the *dorso-laterals*. Then finally come the *carinals*, forming the mid-line of the aboral surface, though these two are absent in some. In the primitive starfishes, the Phanerozonia, the marginals are conspicuous and closely set (it is this fact to which the subclass name, meaning 'visible edge', refers), while in the more advanced forms (sometimes united as a group, Cryptozonia, 'hidden edge') the marginals are no bigger than the other lateral and aboral plates.

Little blisters of the body wall, *papulae*, formed of the ciliated external and coelomic epithelia, with a thin connective tissue layer sandwiched between them, bulge through the spaces between the ossicles, so that, in addition to the oxygen which gets into the water vascular system through the tube-foot walls, a certain amount of gaseous exchange takes place between the sea-water and the perivisceral coelomic fluid; the papulae have a further function to do with excretion (p. 45). In some phanerozones the papulae are restricted to isolated regions, the *papularia*, on the aboral surfaces, usually close to the proximal end of each arm, as in *Pectinaster*.

The whole system of ossicles is flexible, and the shape of the body can be altered by special skeletal muscles. Most important are the upper, lower and lateral transverse ambulacral muscles on the oral side of the arms, which are for altering the depth of the grooves, and the circular and longitudinal muscles over the whole body just below the coelomic epithelium, which are for altering the shape of the body.

The alimentary canal

The mouth, surrounded by an uncalcified peristome stretched across the space within the peristomial ring of ossicles, leads into a vertical gut (Fig. 4). The first part of this, a short oesophagus, may have ten pouches in its walls reminiscent of those of the crinoid gut (not shown in Fig. 4). The next part, the stomach, has a wide *cardiac* portion, held by radial mesenteries to the wall of the disk, followed above by a smaller *pyloric* portion, from which ten *hepatic caeca* arise, two in each arm. Each caecum is heavily sacculated, and it is thought that here digestion is continued and absorption takes place. Apart from the expected mucous and enzyme-secreting cells, the walls of the caeca contain glycogen-storing cells, the contents of which disappear in starved animals[22]. Lastly, there is a short *rectum*, usually with a branching diverticulum, the *rectal sac*. The gut may end blindly in some of the burrowers.

The coelom

The general body cavity is not subdivided as it is in so many other echinoderms; the perivisceral coelom of the disk is a single cavity, continuous with that in the arms. The tubular coelomic systems, water vascular, haemal and perihaemal, are well developed and have probably been studied in more detail in this group than in any other.

The water vascular system

As usual, the main job of this system is to supply the fluid necessary for the hydraulically operated tube-feet. One of the plates on the aboral surface of the disk, derived from the ring of interradial plates formed first of all (p. 131) is pierced by a large number of very small pores, together called the *madreporite*; part of this plate curves towards the interior of the disk to form a cranny in its inside surface, the importance of which we shall see when we consider some of the other tubular coelomic systems. The madreporite leads down into a nearly vertical tube bearing the somewhat unjustified name *stone canal*, because early workers found its walls strengthened with tiny spicules against which their instruments tinkled and grated as they dissected. The fossilized remains of these ossicles have even been found in the earliest asteroids, the Somasteroids[35]. The canal is not a simple tube but usually contains a scroll-shaped projection from one side of its inner wall (Fig. 4,*AA*) to ensure circulation in the tube (towards the mouth inside the flaps of the scroll, in the opposite direction outside). At its oral end the stone canal opens into a *circum-oral water ring*; at its aboral end it is said[27] to open to the surrounding axial sinus and also to have a tiny projection, the *madreporic ampulla*, within the cranny of the madreporic plate. Arising from the wall of the circum-oral vessel are five pairs of interradial *Tiedemann's bodies*, thought to secrete some of the coelomocytes. In some asteroids (though not in the common *Asterias*) the vessel also bears *polian vesicles* interradially, which are muscular sacs with the probable function of maintaining turgor in the system.

From the water ring (internal to the skeleton) arise the radial water canals (external), one to each arm; the canals pass through the skeleton just outside the peristomial ring, and the pores through which they pass can be seen in some early fossils. Each canal passes along its arm above (aboral to) the bridges formed by the lower transverse ambulacral muscles; between these muscles, on alternate sides, branches are given off to the tube-feet, and each tube-foot has an ampulla internal to the skeletal pieces. A valve at the point where the branch from the radial canal joins the lumen of the tube-foot isolates each tube-foot/ampulla unit from the rest of the water vascular system so that the necessary hydrostatic pressure can be built up. The structure and activity of the tube-feet and their ampullae is dealt with more fully in Chapter 9. Each radial canal ends at the arm-tip as a terminal tentacle, basically a tube-foot with no ampulla, whose function is mainly sensory.

One other structure, not really part of the water vascular system, can conveniently be dealt with here—the *dorsal sac* (Fig. 4). This tiny isolated vesicle shares the cranny in the madreporic plate with the madreporic ampulla. Projecting into it is part of the haemal system (see below). Apparently the madreporic ampulla and the dorsal sac pulsate out of phase, to bring about circulation in the aboral parts of the water vascular and haemal systems. The sac has acquired considerable importance, as will be mentioned later, because there is evidence[27] that embryologically it represents the *right anterior coelom*, and this helps to establish the fact that the echinoderm coelom consisted originally of three pairs of sacs, only the left side having its full complement in present-day adults.

Haemal system

As usual, this has a number of ring elements round the gut connected together by an axial part, and radial components leading off into each arm. The *oral haemal ring*, running in a septum in the perihaemal ring (p. 42), gives off the *radial haemal strands*, also in a septum, which pass up each arm external (oral) to the radial water vessels. Round the pyloric stomach is the *pyloric haemal ring*, giving off branches, the *gastric haemal tufts*, to the walls of the cardiac stomach, and four branches, the *hepatic haemal strands*, to the walls of the hepatic caeca of each arm; these gastric parts of the haemal system are not enclosed in perihaemal elements. The *aboral haemal ring*, round the rectum, gives off two branches to each arm, leading to the gonads. In the asteroids it is apparently not possible to distinguish a separate axial haemal system (equivalent to the spongy body of crinoids) from the *axial organ*—the two seem to be one structure. This lies to one side of the stone canal and sends an aboral process, the *head-piece*, into the dorsal sac within the cranny of the madreporic plate. The pulsations of the dorsal sac, mentioned above, and the contained head-piece presumably bring about circulation in the aboral parts of the tubular systems, but why this should be necessary, rather than relying on ciliary action, is not easy to see. Other parts of the haemal system, such as the axial organ, gastric haemal tufts and aboral haemal ring, have also been seen to contract[27].

Perihaemal system

This system surrounds the oral, aboral, radial and axial parts of the haemal system. The *oral perihaemal ring* round the oesophagus sends a *radial perihaemal canal* into each arm; both ring and canals are subdivided by a septum (Fig. 4) so that the ring consists of inner and outer portions, and the canals of two elements side by side. The oral haemal ring and the haemal strands lie within this septum. The radial perihaemal canals, like the haemal strands, are said to give off branches to the tube-feet. From the outer haemal ring arises the *axial sinus*, surrounding axial organ and stone canal. The sinus and the stone-canal communicate aborally. The *aboral perihaemal ring* round the rectum, and, of course, surrounding the aboral haemal ring, is not in communication with the axial sinus but is a separate cavity giving off branches to the gonads.

The nervous system

It is important to remember that in all echinoderms a thin nerve plexus underlies the entire epithelium covering the body; in places it is thickened into fairly definite tracts, though these are anything but separate nerves by vertebrate standards.

The nervous system of the asteroids has received far more attention than that of other echinoderm groups, chiefly through the work of Smith[30, 31, 32, 76] and Kerkut[64, 65]

Originally, as in crinoids, three separate systems were recognized: ectoneural, hyponeural and entoneural, but more recently it has become evident that the hypo- and entoneural systems are continuous, so that a division between these two is inappropriate. It would seem better to divide the whole system into *superficial* and *deep* parts. The superficial part is mainly sensory; it is composed of the general subepithelial plexus and those parts of it which are specially thickened to provide through-conducting paths. The chief of these are, first, the *circumoral nerve ring*; secondly, the *radial* (=*perradial*) *nerve cords* in the mid-line of the underside of each arm; and, lastly, the *adradial nerve cords*, two to each arm, which pass along the underside of each arm lateral to the tube-foot/ampulla canal (Fig. 4). Any special regions of the body, such as special ciliary tracts or aggregations of sensory receptors, normally have a thickening of the plexus just beneath them.

The *deep* component is predominantly motor. It consists of segmentally arranged centres, internal to the superficial plexus, and various nerve tracts to effector organs. The main centres lie adjacent to the tube-foot/ampulla junctions: first, there are *Lange's nerves* (better called *Lange's centres*) just aboral to the lateral parts of the radial nerve cord. These are aggregations of twenty or more neurones lying in the floor of the radial perihaemal canals, separated from the nerve cords below by a thin layer of connective tissue across which nerve fibres pass to join the two systems. Nerves pass from Lange's centres to the inferior transverse ambulacral muscles just above them, and to the tube-foot/ampulla system. Secondly, there are motor centres lateral to the tube-feet, innervating the ampullae (which thus have a double innervation) and in addition sending lateral motor nerves up the inside of the body wall just below the coelomic epithelium. These nerves supply various skeletal muscles. In those forms with two lobes to each tube-foot ampulla, such as the phanerozone *Astropecten*, the lateral motor centres innervate the lateral lobes while Lange's centres innervate the medial lobes.

Besides the general scattering of neurosensory cells over the asteroid body, there are five light-sensitive *optic cushions*, one at the base of each terminal tentacle. Each cushion contains numerous ocelli in the form of cups, the external part of which is sometimes modified into a lens. The walls of the cup consist of cells containing a red pigment and, interspersed between them, retinal cells with nerve fibres passing into the nearby radial nerve cord.

The ambulacrum

From what has been said about the radial parts of the tubular coelomic systems, it can be seen that they all lie *outside* the skeletal pieces of the ambulacrum, though of course covered by an external epithelium. This is termed an *open ambulacrum*, and is the type first seen in the Pelmatozoa and the extinct Eleutherozoa. In the other three living eleutherozoan classes, however, the skeletal pieces close over the radial elements of the various systems as further protection, thus forming a *closed ambulacrum*. The nature of the ambulacrum has been used to support phyletic relationship[7] but it will become evident later (p. 170) that the echinoderm ambulacrum has closed over independently at least twice.

Some aspects of asteroid physiology

Respiration. It has already been said that oxygen from the sea-water is taken into the body via tube-feet and papulae. In some phanerozone starfishes special aggregations of papulae are present in grooves between some or all of the marginal plates of the arms. These are called *cribriform organs*. Cilia drive water in an oral direction through these organs, and it appears that the strong currents are also used to bring food into the ambulacral grooves from the dorsal (aboral) surface. In other phanerozones a respiratory chamber may be present on the aboral surface: special spines called paxillae (more fully described in Chapter 8) can close over to lie parallel to the aboral body wall and abut against their neighbours to form a chamber between them and the body. Papulae lie in the floor of this chamber, and water, drawn in and out by a combination of muscular contractions of the body wall every ten to twenty seconds and ciliary action, passes over them for gaseous exchange. The chamber is also used by some starfishes for brooding the larvae.

Excretion. Like other echinoderms, the asteroids lack a definite excretory system; but injection of coloured particles has revealed that these, and presumably other waste products too, are ingested by coelomocytes, some of which pass into the papulae and are extruded by rupture of the papular wall, while others apparently pass through the external epithelium of certain parts of the body, particularly the disks of the tube-feet, to be extruded as slime.

Osmotic regulation. There seems to be none: the body fluids are almost isotonic with sea-water. The ionic tolerance varies astonishingly, some starfishes being intolerant to salt level change, while others enjoy much greater freedom, and hence ecological range (e.g. *Asterias rubens* which ranges from the North Sea at 30‰ salinity to the Baltic at 8‰)[23, 24].

Feeding. The primitive asteroids, as one would expect from their forbears, were probably ciliary feeders, relying on currents over the animal's surface to collect food particles, passing them in strings of mucus to the ambulacral grooves and thence to the mouth; other particles would probably be picked up from the surface over which the animal moved, possibly by the tube-feet, and added to the same stream. Then the tube-feet acquired suckers, probably for better locomotion in the first place; and later the ability to prey on bivalves was seized upon. Considerable interest has been focused on the problem of how starfishes open bivalves. It has been shown recently[29] that the combined pull of the starfish tube-feet and body muscles on a bivalve's adductor muscle may be as high as 3,000 gm. It is commonly said that the starfish wins the tug-of-war because it can keep up a persistent pull of this magnitude for hours on end. This does not seem to be the case: between five and ten minutes is all the time required for the starfish to pull the shells open about 0·1 mm, which is sufficient to insinuate the stomach and start digestion; and the force is relaxed and reapplied at intervals until the adductor muscle has been rendered ineffective.

The evolution and adaptive radiation of the Asteroidea

The views expressed here are contrary to those[7] giving the ophiuroids a closer phyletic connexion with the echinoids than to the asteroids. There would seem to be excellent grounds, particularly from palaeontological evidence, for regarding the group Stelleroidea (asteroids and ophiuroids) as natural, and it is logical to regard it as a class within the Echinodermata at the same taxonomic level as the Crinoidea, Echinoidea and Holothuroidea. However, so deeply engrained in common usage are the terms Asteroidea and Ophiuroidea, and so comparatively distinct are the living members, that it would seem inappropriate to relegate them to the rank of sub-class, which is the most recent suggestion of Spencer[35] and which is followed by Ubaghs in the French *Traité de Paléontologie*[111]. As Spencer points out, the earliest starfishes show characters consistent with their being considered ancestral to both the later starfishes and the ophiuroids, and he suggests including these early forms in a separate sub-class, the Somasteroidea. Like Hyman[7], I regard the somasteroids as much closer to the asteroids than to the ophiuroids, and therefore I treat them here as a sub-class of the Asteroidea, but it should be made clear that some palaeontologists consider the three groups of star-shaped echinoderms as of equal status.

The earliest somasteroids are *Villebrunaster* (named after its collector, Villebrun) and *Chinianaster* (after the locality) from the Upper Tremadoc and Lower Arenig beds of the Lower Ordovician of St Chinian, in the south of France[33, 34]. The specimens were found embedded in flint-like nodules lying in the soil of vineyards. The owners of the vineyards generally cleared the nodules from the soil for convenience of cultivation and left them in spoil-heaps at the edge, a rich source for the fossil collector of all sorts of animal remains. The third somasteroid is *Archegonaster*, from the Upper Arenig beds of the Lower Ordovician (slightly later than the other two) of Bohemia. Specimens of this form were also preserved in nodules. An interesting feature of preservation, which also seems to throw light on the behaviour of the animals when they lived, is the fact that generally only the disks and proximal parts of the arms are found in the nodule, and very often the arms are flexed away from the oral surface. Spencer[33, 35] suggests that the animals were burrowers, and that they lived in a manner similar to many recent asteroids and ophiuroids, that is, with only the arm tips protruding above the surface of the substratum; then, only those ossicles which were actually buried at death were held together sufficiently long for fossilization to occur. In this case preservation occurred by the precipitation of minerals round the animal. Subsequently the calcite of the skeleton was dissolved away, so that most of the studies have been made on casts; nevertheless, the impressions left behind contain fine detail of spines, joints and muscle facets, in fact, almost as much as could be hoped for from a recent asteroid skeleton. So our knowledge of this basal group is surprisingly full.

The most primitive somasteroid, as well as one of the earliest, appears to be *Villebrunaster* (Fig. 5*a*, *b*). This form has a large disk with a wide pentagonal peristome and five petaloid arms. Along the midline of each arm on the oral surface are two columns of alternating ambulacral ossicles, each shaped like a half-cylinder, which do not contain a groove between them. Instead, a cylindrical channel, presumably the site of the radial water vascular canal, runs between the two columns, totally enclosed. In this feature they are unlike the true asteroids, but, as we shall see, the channel tends to become open on the oral face in later forms. Branches from the enclosed canal passed out on alternate sides to deep concavities in the oral faces of each ambulacral ossicle. It appears highly probable that each concavity was the site of insertion of a tube-foot with a small ampulla which did not lie to the inside of the skeleton.

It should be mentioned that in no known early (pre-Devonian) asteroid are there pores between the ossicles for the passage of the canal to an ampulla. The ambulacral columns diverge at their inner ends to become the edge of the peristome; the proximal member of each column, which comes to abut against its neighbour from the next arm at the adjacent interradius, is slightly larger than the rest of the series, paralleling the enlarged *ad*ambulacrals which occupy the same position in the mouth region of later, true, starfishes. The rest of the oral skeleton consists of many longitudinal series of elongated ossicles, the *virgalia*, arranged at an angle to the ambulacrum of each arm. The aboral skeleton consists of small triradiate spicules in the form of a meshwork.

In the somewhat less primitive but contemporaneous somasteroid *Chinianaster* the arms are not petaloid in shape. The main anatomical advances towards the true asteroid condition are, first, that the outermost virgalia of

---

Fig. 5 EVOLUTION AND ADAPTIVE RADIATION IN THE ASTEROIDS

a The first asteroid, a somasteroid *Villebrunaster* (L. Ord.); oral view of skeleton.

b to *h*, diagrammatic transverse sections of the arms of a fossil series of asteroids showing the probable evolution of the ossicles.

b *Villebrunaster* (L. Ord.).
c *Chinianaster* (L. Ord.). The outermost virgalia have become the marginals.
d *Archegonaster* (L. Ord.). The innermost virgalia have become the adambulacrals.
e *Urasterella*, a hemizonid (U. Ord.). There are no virgalia.
f *Petraster*, the first true asteroid (L. Ord.).
g *Hudsonaster* (Ord.).
h *Xenaster* (Dev.). Pores between ambulacrals and adambulacrals, probably for canals to ampullae.

i to *l*, typical British asteroids, showing the main trends.
i *Astropecten*, a phanerozone which burrows in sand or gravel.
j *Anseropoda*, a spinulosan with thin body, also a burrower.
k *Asterias*, a forcipulate which lives on the surface.
l *Solaster*, a spinulosan.

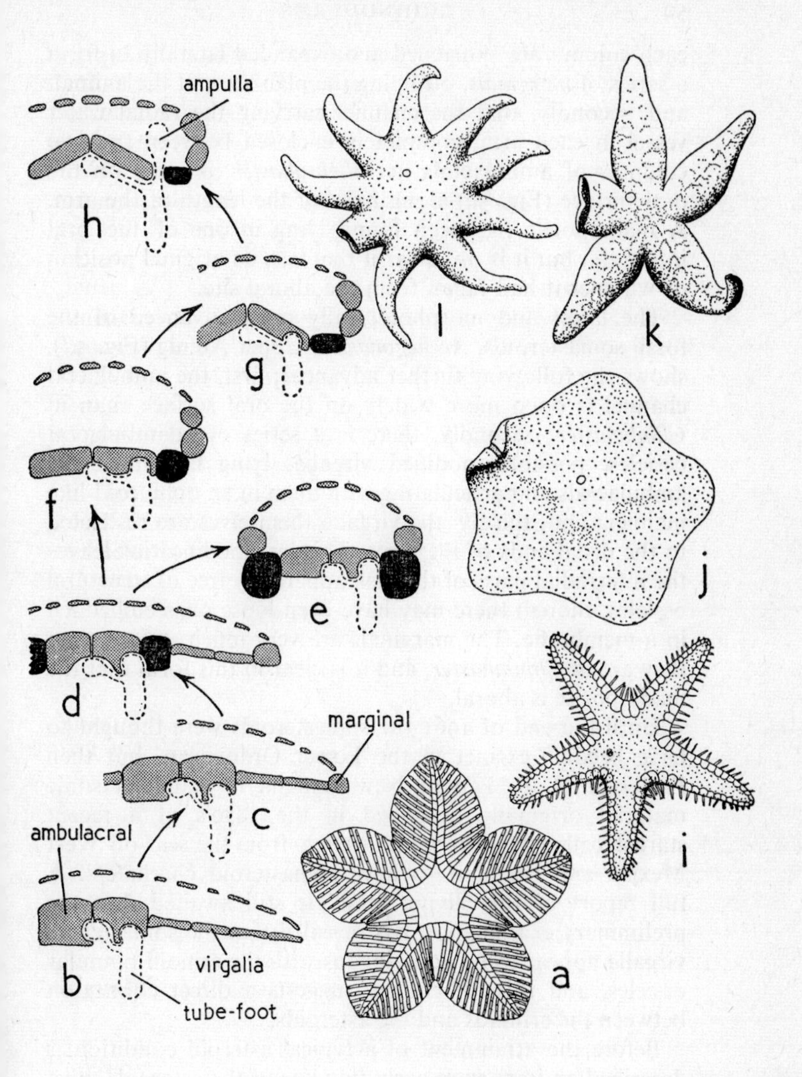
ampulla
h
g
f
e
d
marginal
c
ambulacral
b
virgalia
tube-foot
a
l
k
j
i

each column are shortened and expanded laterally to form a series of *marginals*, outlining the plan-view of the animal, and, secondly, that the channel carrying the radial water vessel in each arm, completely enclosed between the two columns of ambulacrals in *Villebrunaster*, is open on the oral surface (Fig. 5*c*) along most of the length of the arm. A madreporite has been found lying in one of the oral interradii, but it is not clear if that was its original position or whether it had fallen from the aboral site.

The latest and morphologically most advanced of the fossil somasteroids, *Archegonaster*, Upper Arenig (Fig. 5*d*), shows the following further advances: first, the ambulacral channel is open more widely on the oral surface than in *Chinianaster*; secondly, there is a series of adambulacral ossicles, possibly modified virgalia, lying lateral to the ambulacrals and articulating with them in an ophiuroid-like manner; and, thirdly, the virgalia themselves are restricted to the distal ends of the arms. This last modification leaves the interradial parts of the disk apparently free of structural ossicles, though there may have been loose ones embedded in a membrane. The marginals are very much stouter than they are in *Chinianaster*, and it is clear in this form that the madreporite is aboral.

Until the end of 1961 the somasteroids were thought to have become extinct in the Lower Ordovician, but then Professor H. B. Fell of New Zealand re-examined some material originally described in the 1870's of a recent starfish called *Platasterias latiradiata* from the seas off West Mexico and found it to have somasteroid characters. A full report of this 'living fossil' is still awaited, but the preliminary examination has revealed that the somasteroid virgalia appear to be homologous with the crinoid pinnular ossicles, and this, of course, suggests a direct connexion between the crinoids and the asteroids.

Before the attainment of a typical asteroid condition is described an important early side-line of the asteroids must be mentioned. This is a group regarded by Spencer[35] as belonging to a separate order, the Hemizonida, and includes forms such as *Urasterella* (Fig. 5*e*) and *Cnemidactis* from the Upper Ordovician, that is, later than the three fossil somasteroids mentioned above. The main feature of the group is that its members have very small disks but long, thin arms. The ambulacral ossicles form the roof of a groove, the sides of which are formed by the large adambulacrals. These occupy most of the rest of the oral surface, displacing the marginals so that they lie on the aboral side of the margin. There are no virgalia. The group possibly extends up to the Chalk (Upper Cretaceous), where the starfish *Arthraster* has similar features, though these may be convergent.

We will now briefly follow the evolution of the true asteroids from the later somasteroids. As in the Hemizonida, the virgalia are lost or transformed completely, and we see now the attainment of a second row of marginals on the aboral side of the edge, the *supramarginals* (the original marginals now being called the *inframarginals*). We also see a feature which parallels a change seen in the hemizonids: the ambulacral ossicles expand laterally from being barely more than half-cylinders in the somasteroids to the true asteroid condition of elongation at right angles to the long axis of the ambulacrum. The concavities for the tube-feet and their ampullae retain the somasteroid condition by lying between two plates rather than in the centre of each plate. In the post-somasteroid starfish the madreporite may be aboral, marginal or oral in position in closely related forms, so this feature does not have any phyletic significance.

The first true asteroid is *Petraster* (= *Uranaster*) (Fig. 5*f*) from the Lower Arenig beds of the Lower Ordovician. This is roughly contemporaneous with the somasteroid *Chinianaster*, but there is no doubt that *Petraster* is a true phanerozone starfish. Among the various species of this genus one can trace the gradual elaboration of the supramarginals from a rather vague series of ossicles close to but not touching the inframarginals along part of the arm to a definite series articulating with the inframarginals all along it[33]. The ambulacral plates are rather poorly known, but there already seems a tendency for lateral expansion. The ambulacral groove is not yet well marked.

In the next stage, typified by *Hudsonaster*, Ordovician (Fig. 5*g*), the main advances are the enlargement of the two rows of marginals into really strong elements for support of the body, the flexure of the ambulacral ossicles to form the typical asteroid ambulacral groove, and the strong attachment of the adambulacrals to the ambulacrals, as in later asteroids. Another interesting point is that in *Hudsonaster* the plates of the apical disk (in the centre of the aboral surface) show the typical echinoderm condition of a central plate surrounded by a ring of five interradial *genital* plates, one of which is the madreporite. Such an arrangement is seldom found in recent asteroids; in those forms which do have a pentameric arrangement of apical plates, such as *Tosia*, there is generally another ring of radial plates inserted between the central and the genitals. However, many starfishes (e.g. *Asterina*) are known to pass through the hudsonasterid condition when the aboral plates are first formed. Finally, in *Hudsonaster* one can see the adoption of the typical later asteroid condition in the mouth-frame: the proximal adambulacral plates of each column are larger than the rest and form the typical mouth-angle plates in each interradius round the mouth, thus paralleling the condition in the somasteroids, where the 'mouth-angle plates' were formed by enlarged ambulacrals.

In *Hudsonaster* we have reached a condition of the skeleton more or less common to all subsequent asteroids. Clearly, the last-remaining feature typical of recent asteroids is the adoption of endothecal ampullae, indicated in the fossils by the presence of pores to the interior between the ambulacral plates. The first signs of this advance are seen in the Devonian *Xenaster* (Fig. 5*h*), in which the pores enter the interior of the arm at the lateral margins of the ambulacral plates, that is, where the ambulacrals and adambulacrals abut. In subsequent palaeozoic forms the tendency is for the pores to shift to a position about halfway along the ambulacral plates and at the same time for the ambulacral plates to oppose rather than alternate. This, the typical phanerozone condition, has persisted to the present day.

It is clear from what has been said that of the living forms the phanerozones are the most primitive. Indeed, all pre-Tertiary forms show the conspicuous marginals characteristic of this group; then, in the Tertiary, we get the first appearance of *cryptozone* starfishes, with less distinct marginals. The great naturalist Percy Sladen introduced the terms Phanerozonia and Cryptozonia as taxonomic orders, and this arrangement is still followed by some authorities[4], but there appears to be good palaeontological evidence that a cryptozone condition originated from the phanerozone stock several times, so there would seem to be grounds for following another classification based on one originally suggested by the Frenchman Perrier at the end of the last century. This grouping recognizes three orders: Phanerozonia, Spinulosa and Forcipulata.

Many of the PHANEROZONIA of recent seas are burrowers. The most typical British forms are *Astropecten* (Fig. 5*i*) and *Luidia*, of the sub-order Paxillosa, which normally burrow in sand or fine gravel, maintaining a connexion between the burrow and the surface of the substratum at the arm tips and sometimes at the disk as well. The papulae are restricted to the oral surface, where they can be protected and kept clear of the falling sediment by the paxillae, the universal presence of which gives the sub-order its name. There are no suckers on the tube-feet. In some members of the sub-order Valvata the general pattern is rather similar to that of the Paxillosa, and one can surmise that the two groups probably had a phyletic connexion; but the British valvatid *Porania* appears to lie close to the transition between the Phanerozonia and the Spinulosa. It is a non-burrower, and the tube-feet have suckered disks for moving over the substratum; there is a thick aboral membrane for protection instead of the paxillate condition normally associated with the phanerozones.

The order SPINULOSA is not sharply separable from the Phanerozonia. The order gets its name from the fact that aboral spines are normally present, usually arranged in groups either borne on a stalk ('pseudopaxillae'), as in the British form *Solaster* (Solasteridae) (Fig. 5*l*), or sessile on the surface, as in *Henricia* (Echinasteridae). The tube-feet of most spinulose starfishes are suckered. There are two British members of the spinulose family Asterinidae, *Asterina* and *Anseropoda* (= *Palmipes*) (Fig. 4*j*), and they show contrasting modes of life. *Asterina*, the small Gibbous Starlet, is a common shore animal, usually found in rocky areas. It is negatively geotactic, and presumably climbs to get closer to the more highly oxygenated surface waters. It is probable that the pull of the body on the tube-feet is the means of perception, because the effect can be reversed by tying a string round the body and pulling upwards. In *Anseropoda* the body is wafer-thin and the outline almost pentagonal, and normal movement is in the opposite direction—it is a burrower. It is peculiar as a burrowing form in that it possesses suckered tube-feet, and not the more usual pointed mucus-plasterers typical of burrowing phanerozones. Burrowing is apparently effected by musculo-skeletal methods rather than by digging with spines and tube-feet; the animal thrusts one side of its body beneath the surface of the substratum by pushing with the other. The animal is characteristic of shell-gravel, so the suckered tube-feet would be well suited to move the particles.

The last order, FORCIPULATA, is characterized by the presence of special pedicellariae (p. 100) with a basal piece and two valves (Fig. 14*g*, *h*). Here belong the asteriids (Fig. 5*k*) and the American Sunstar, *Heliaster* (not to be confused with the British spinulose form *Solaster*). There is also the very peculiar deep-water form *Brisinga*, originally considered primitive and possibly close to the ophiuroids because of its possession of a distinct disk and long spiny arms, and, even more ophiuroid-like, very small ampullae to its tube-feet, and these partially embedded in the ossicles of the oral side. Perrier regards the Brisingidae as the most primitive asteroids, because of their ophiuroid-like nature, but Ludwig, Sladen and others consider that this group and the ophiuroids are convergent.

THE OPHIUROIDEA

THE members of this class, with much less variety of external form than any other, are delimited by their small disks and long spiny arms, though there is one exception (p. 54). It is these arms to which the class name refers ('snake-like'); the ease by which they break when handled has led to their common name 'brittle-stars'. Photographs of the sea-bottom often show huge aggregations of these animals, sometimes overlapping and even several deep, and this has given rise to the suggestion that they are the most successful of the echinoderm classes, judged on the criterion of numbers. Unlike most of the asteroids, the ophiuroids do not use only their tube-feet for locomotion, but usually move by sinuous flexures of the arms, one arm and the disk being thrust forward by oar-like movements of the two other pairs.

General body plan

The whole of the alimentary canal, the axial parts of the tubular coelomic systems and the gonads are contained in the disk (Fig. 6). The arms consist of little more than ossicles, their operating muscles and the radial coelomic components. There is no anus, the star-shaped mouth on the oral surface serving for the removal of undigested food-remains. Primitively, the arrangement of plates on the aboral surface of the disk is a *central* plate, surrounded by a ring of five *radials* (Fig. 12*c*). More cycles of plates, usually multiples of five, then surround this central pattern to cover the disk. In more advanced ophiuroids this rather regular pattern of ossicles is obscured by reduction in size of the original plates and the addition of other small plates. A pair of conspicuous plates, the *radial shields*, used extensively in taxonomy, lie close to the origin of each arm. On the oral side five interradial *oral shields* lie in the angles of the arms. These plates originate in the embryo as the second whorl on the aboral surface, that is, outside the ring of radials, and migrate to the oral surface; one of them is the madreporite, so they are most likely homologous with the basals of asteroids and the genitals (= basals) of echinoids. At the base of each arm there are a pair of *jaw plates* and a pair of *aboral shields*, homologous with the ambulacral and adambulacral plates respectively of the peristomial ring in asteroids.

The arms continue *inside* the disk on the oral surface. At their bases on each side is a slit, leading to a *genital bursa*, a sac which bulges up into the disk, the function of which is both respiratory and reproductive (p. 130). Each slit is bordered by a *bursal plate*. The main plates of the arms are a longitudinal series of ossicles called *vertebrae*, formed by fusion of two plates homologous with the asteroid ambulacral plates. Each hinges against its neighbour so that the arm can bend in a horizontal plane; movement is brought about by two pairs of muscles, the *oral* and *aboral intervertebrals*. Surrounding the vertebrae are four series of plates—the *orals* in the mid-ventral line, the *aborals* mid-dorsally, and the *laterals* which bear spines. The tube-feet emerge from between the orals and laterals, and special modified spines on the laterals, the *tentacle scales*, lie at their bases.

Fig. 6 BASIC ANATOMY OF AN OPHIUROID

Diagrammatic vertical section through the disk and part of one arm of a brittle-star, passing through one genital bursa, one gonad, the axial complex and one ambulacrum. On the right, transverse section of an arm, passing through a pair of tube-feet. See Fig. 2 for key to systems.

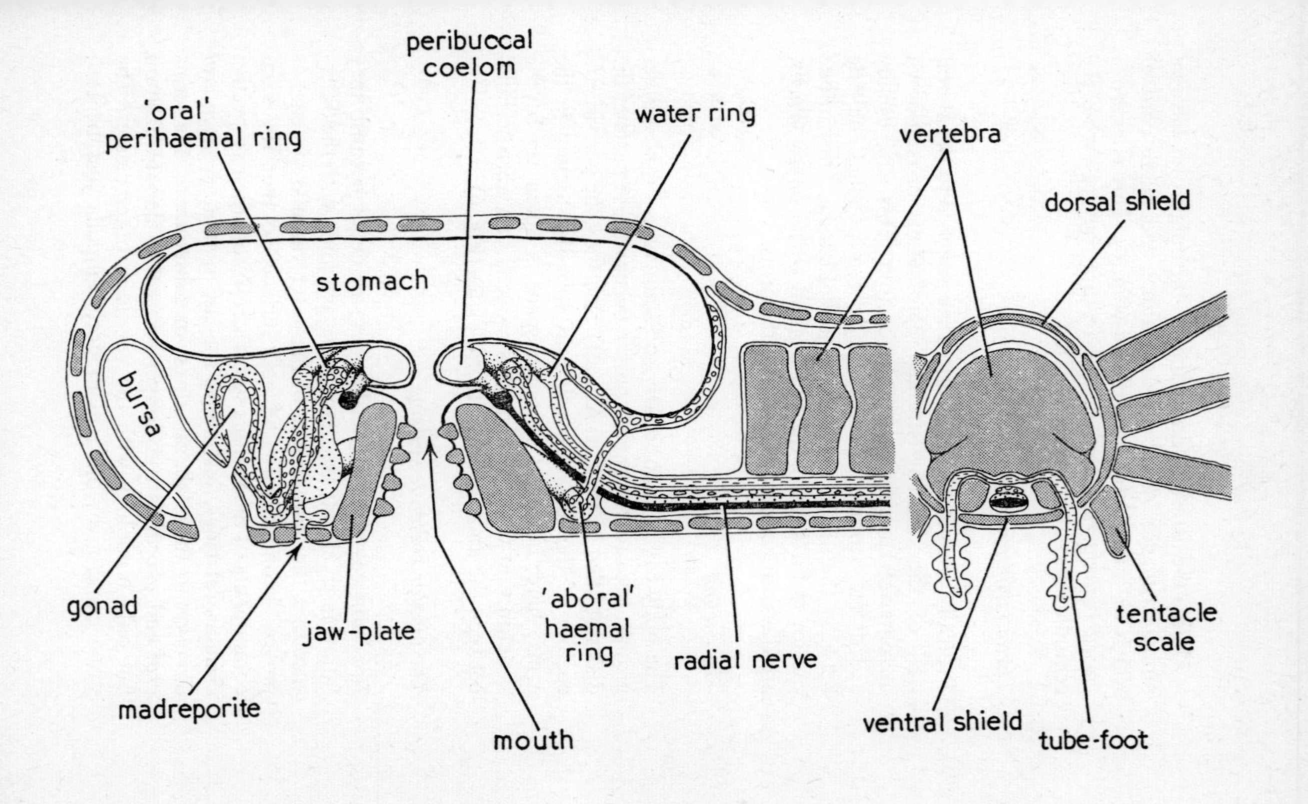
peribuccal coelom
'oral' perihaemal ring
water ring
vertebra
dorsal shield
stomach
bursa
gonad
jaw-plate
madreporite
mouth
'aboral' haemal ring
radial nerve
ventral shield
tube-foot
tentacle scale

Unlike the asteroids and crinoids, the ophiuroids have an ambulacrum the radial coelomic tubes and nerve cord of which lie internal to the skeleton (Fig. 6). This is called a *closed ambulacrum*, and in this feature they resemble the echinoids and holothuroids (p. 84).

The alimentary canal

Typically, the mouth opens into a sac attached to the aboral wall of the disk by strands representing broken-down mesenteries running across the coelom. This sac, usually called the stomach, has typically ten pouches bulging between the inside walls of the genital bursae. Only in one species so far known, the sub-tropical *Ophiocanops fugiens*, are there any continuations into the arms.

The coelom

The body cavity is not nearly so extensive as in most other groups, because the genital bursae occupy those parts of the disk not filled by the gut. As in the asteroids, the perivisceral coelom of the disk is continuous with the coelom in the arms; here, however, so much space is taken up by the skeleton that the arm coelom is reduced to a narrow canal between the aboral and vertebral plates (Fig. 6).

The tubular coelomic systems

It is unnecessary to describe these in detail because they follow fairly closely the plan for the asteroids, with the difference that the madreporite of the water vascular system has moved secondarily to the oral surface, so that the axial components of the various systems run from their respective circum-oral rings, just above the jaw apparatus, *in an oral direction* to their 'aboral' rings; in consequence, the terms *oral* and *aboral* can no longer be used to denote position but only homology with the vessels in other classes. The stone canal gives off a madreporic ampulla just beneath the madreporic plate; the oral water ring, in addition to the five radial water canals, gives off four or five polian vesicles and ten branches which subdivide to become the lumina of the twenty buccal tube-feet, only ten of which are usually visible externally, the other ten arising close to the mouth. The ordinary tube-feet of the arms lie side by side, not alternately. In many Eurylae (p.64) the axial organ of the haemal system consists of a lighter coloured 'oral' part and a darker 'aboral', the latter possibly homologous with the asteroid head-piece. The axial sinus of the perihaemal system is divided by the axial organ and stone canal into left and right parts, the left corresponding to the asteroid axial sinus and the right to the asteroid dorsal sac. Again as in asteroids, the sinus communicates 'aborally' with the stone canal.

Nervous system

The main nerve ring lies just oral to the axial sinus and gives off branches to each of the buccal tube-feet, which are sensory, and to each arm. These branches pass along the oral side of the vertebral ossicles either in a groove or in a canal.

The origin of the ophiuroids

The problem of the origin of the ophiuroids is an outstanding example of how easy it is to draw conflicting pictures of phyletic relationship using evidence from different disciplines. It is probably true to say that if no fossil evidence were available a connexion between the ophiuroids and the echinoids would be most attractive, the similarity of body plan between ophiuroids and asteroids being purely convergent. Let us look briefly at the features in which the ophiuroids resemble the echinoids. First, there is the nature of the ambulacrum, a 'closed' system in both (p. 84), with the consequent presence of epineural canals in each radius and round the mouth; secondly, the tube-feet of both groups pierce the ambulacral plates rather than pass between them in the asteroid manner; thirdly, both have pluteus larvae; and, lastly, there is some evidence of biochemical affinity: echinoids and ophiuroids are said to share the same type of sterol (cholesterol) while asteroids have a different type (stellasterol)[115]. It must be admitted that this is a pretty convincing set of similarities. Yet there are grounds for mistrusting them, even disregarding the fossil picture. For instance, it is shown (p. 130) that larval convergence has taken place in unrelated groups in answer to particular larval conditions, so that little weight can properly be placed on this feature. Again, it is becoming increasingly clear that the distribution of sterols is not nearly so clear-cut as originally thought[121], so the biochemical evidence, too, may be unreliable as a guide to phylogeny.

But, luckily, considerable fossil evidence is available, and this shows that without much doubt there is true phyletic relationship between asteroids and ophiuroids. The line from the somasteroid *Villebrunaster* through *Pradesura*, *Palaeura* and *Stenaster* in the Ordovician to *Lapworthura* in the Silurian shows about as much of the establishment of those characters typical of present-day ophiuroids as one could hope to get from the decidedly fragmentary fossil record of the early Palaeozoic. The series shows the gradual increase in size of the asteroid-like ambulacral plates until they become the vertebral ossicles of the ophiuroids; it shows the gradual envelopment of the radial water vessel until it becomes embedded in the substance of the vertebrae, and the gradual enlargement and migration of the adambulacrals until they become the lateral arm shields; round the mouth the transition from the asteroid mouth angle plates to the ophiuroid jaws can be followed, and from the asteroid first adambulacral to the ophiuroid adoral (lateral buccal) shields; and the madreporite in some of the early asteroids is oral, as it is in ophiuroids. Surely such a wealth of evidence can hardly fail to be convincing. Let us, then, look at the evidence of ophiuroid origin in more detail.

Judged on the criteria of general body form (the separation

of disk and arms), the oral position of the madreporite and the ossicle arrangement in the arms, the first recognizable ophiuroid to appear is *Pradesura*[35] from the Lower Arenig beds of the Lower Ordovician (Fig. 7*a*, *b*). The oral surface of each arm has three types of plate: first, a double row of typical somasteroid-like alternating *ambulacrals* bordering a central channel for the radial water vessel and containing the typical hollows for tube-feet and their incipient ampullae shared between two adjacent plates: secondly, a column of longitudinally elongated *sub-laterals* lateral to each ambulacral column; and, thirdly, a similar column of rather larger *laterals* outside the sub-laterals, each of which bears a column of spines on a ridge set at right angles to the long axis of the arm, that is, in the typical ophiuroid manner. The ambulacrals closely resemble the homologous plates in the earliest somasteroid starfish *Villebrunaster* (p. 47) which is roughly contemporaneous with it, and so it is likely that both laterals and sub-laterals have evolved from the somasteroid virgalia, but which of the two columns is a homologue of the *lateral shields* of later ophiuroids is not at all clear; some authorities[9] suggest that the shields are formed by the fusion of both series. The mouth frame closely resembles that of the early somasteroids. The disk skeleton consists of small imbricating scale-like plates on both sides.

The next forms to appear are *Eophiura* and *Palaeura*, both from the Upper Arenig beds. The main advances over *Pradesura* are the lateral expansion of both lateral and sub-lateral plates and the gradual change, in *Palaeura* at least, from the early arrangement in which the tube-feet bases are shared more or less equally between two adjacent ambulacrals to the more ophiuroid-like condition in which far more of each basin lies on one plate.

There is little doubt that these three fossils form a series representing the earliest experiment into an ophiuroid type. Spencer[35] places them in an order on their own, STENURIDA. In all the stenurid ophiuroids only the disk and bases of the arms are known, probably because they were burrowers, and, like some recent stelleroids, sat in the substratum with only the arm-tips exposed above the surface; at death, this left the disks and arm-bases more favourably placed for preservation.

An early offshoot from the stenurid forms was the order AULUROIDEA, the principal genus of which is *Aspidosoma* (Fig. 7*c*). In this the ambulacral ossicles are alternating, but the later ophiuroid condition, in which the water vascular canals leading to the tube-feet and their ampullae pass through the plates rather than lie on the oral side of them, is anticipated. Another typical ophiuroid feature is the enlargement of the adambulacrals to occupy the whole of the lateral faces of the arms, so that they are now known as *lateral shields*; they bear typical ophiuroid spines projecting laterally.

The next group, regarded by Spencer[35] as an order, OPHIURIDA, comprises forms which have the usual ophiuroid characters of disk delimitation and oral madreporite but differ from the stenurids and auluroids in having opposing rather than alternating ambulacral plates. This is a change important to the ophiuroids, because clearly it has led to the possibility of fusion of the two members of a pair to form the typical *vertebrae* of modern forms. Ophiuroids like *Stenaster* (Fig. 7*d*) and *Taeniaster* (both from the Middle and Upper Ordovician) are the first of this group to appear. In *Stenaster* we see that the ambulacrals are already tending to move inwards, while the large adambulacrals are tending to surround them. The ambulacral channel, which in life held the radial water canal, is closed, and the ambulacral plates articulate by means of ball-and-socket joints. The basins for the tube-feet and their ampullae are still shared between two adjacent ambulacral ossicles.

The final stage in the attainment of a pattern recognizable as typically ophiuroid is seen in the Silurian form *Lapworthura*[109, 110] (Fig. 7*e*). Advances on the *Stenaster* plan are clearly seen in the following characters: first, the branch of the water canal leading to each tube-foot was enclosed in the substance of the ambulacral ossicle; secondly, the

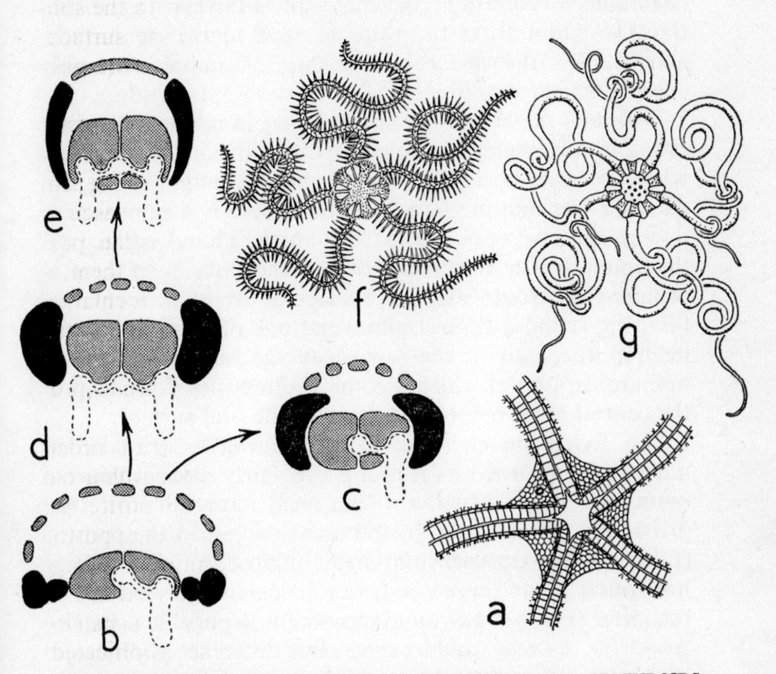

Fig. 7 EVOLUTION AND ADAPTIVE RADIATION IN THE OPHIUROIDS

a Pradesura (L. Ord.), the first ophiuroid. Only the disk and the proximal parts of the arms are known.

b to *e*, diagrammatic transverse sections of the arms of a series of early ophiuroids, showing the probable evolution of the ossicles.

b Pradesura. The laterals and sub-laterals have evolved from the somasteroid adambulacrals (see Fig. 5*d*).

c Aspidosoma (Dev.), an auluroid. The fused laterals are enlarged.

d Stenaster (Ord.). Ambulacrals moving inwards.

e Lapworthura (Sil.). Compare with transverse section in Fig. 6.

f Ophiocomina, a typical recent ophiurid with arm flexure in the horizontal plane only.

g Astroschema, a typical euryalid with flexure in all planes.

adambulacrals are thin, and thus show a further stage in the transformation from the asteroid type to the ophiuroid; and, thirdly, the surfaces of contact of ambulacrals and adambulacrals are reduced, as in modern ophiuroids.

With this remarkably complete series in mind, it is interesting to look again at the embryological evidence and see what else can be considered of phyletic importance. To mention just one point, we see that there is a remarkable resemblance between the ossicles on the aboral surface of the young brittle-star and those on the same surface of a primitive asteroid: the interradials of asteroids form the first ring round a central plate, and one of them holds the madreporite, and in the same way the buccal shields of ophiuroids, one of which has the madreporite, form round the central plate before migrating to the oral surface.

The living members of the Ophiuroidea are almost universally regarded as forming two fairly distinct orders, OPHIURAE and EURYALAE. The main criterion for this division is the structure of the arm ossicles: in the former (Fig. 7*f*) they restrict movement almost entirely to the horizontal plane (*zygospondylous* articulation), whereas in the latter (Fig. 7*g*) all-round movement is possible (*streptospondyly*). As one would expect, this difference is reflected in broad differences in the biology of the two groups: whereas most of the ophiuroids move in or on the surface of the substratum, many euryalids are capable of clinging to objects with their arms, so that they can climb through the fronds of algae, etc. For instance, the euryalid *Asteronyx* is said to move over beds of pennatulids (sea-pens, of the phylum Coelenterata) feeding on their polyps. During their development some euryalids are said to pass through a zygospondylous condition, suggesting that the Ophiurae are more primitive.

5

THE ECHINOIDEA

IN THE sea-urchins, alone among the living echinoderms, the skeletal elements form a rigid theca or *test*. Who can fail to be struck by the symmetrical beauty of these objects when cleaned by the sea and cast up on the beach? In the more primitive regular forms the smoothness of the tests is not broken by food grooves, but these may be secondarily re-formed in the more advanced irregulars. The irregulars have secondary bilateral symmetry superimposed on the basic radial plan, giving them distinct anterior and posterior ends, so that they move in one direction only, and conferring on them the consequent advantages, well exploited, in colonizing habitats forbidden to the regulars.

General body plan

The test is composed of interlocking plates radiating in rows from the apex round to the mouth in the centre of the under (oral) surface. The plates of all living echinoids are in alternating double columns, ambulacra and interambulacra; in some extinct groups the number of columns of each sort range from one to twenty or so, but all modern echinoids, regular and irregular, have two of each. The ambulacral plates are distinguished by having a pair of pores through which pass the canals leading to the tube-feet. The echinoids have a *closed ambulacrum*, so that the radial water vessels and the other radial structures are all internal to the skeleton.

In the urchin immediately after metamorphosis the whole of the aboral surface is covered by an apical disk of plates, consisting of a central *suranal*, through which the anus opens, a ring of five *basals* (more often called *genitals*) through which the gonoducts open, and between these a second ring of five *radials* (sometimes called *oculars*, because the terminal tentacles which they bear are often heavily pigmented and were at one time thought to be light-sensitive). One of the basal plates is perforated by many pores and is the madreporite. The outer edges of the radials are the growing points for the double columns of ambulacral plates, and the outer edges of the basals do the same for the interambulacrals. There is no equivalent of the terminal plates of asteroids and ophiuroids, unless the radials are to be considered homologous with them on the grounds that they bear the terminal tentacles of the ambulacra.

As more and more plates are budded off and the test grows in volume, the apical disk remains roughly the same size and stays in the centre of the aboral surface. Changes occur in it in the irregulars, but these are beyond the scope of this book. Round the mouth is a strong *perignathic girdle* (Fig. 8), consisting of modified ambulacral and interambulacral plates. Each terminal pair of ambulacral plates forms an arch-like *auricle*, while each interambulacral pair forms a solid *apophysis*. This girdle is the main attachment for the fascinating masticatory apparatus, *Aristotle's lantern* (described below), possessed by all regulars and some irregulars. Stretched across the space bounded by the girdle is a flexible peristomial membrane, containing embedded plates.

Fig. 8 BASIC ANATOMY OF AN ECHINOID

Diagrammatic vertical section through the body of a regular echinoid, based on *Echinus*. The section is taken through one ambulacrum on the right, of which three tube-feet and their ampullae are shown. On the left is a section through the axial complex.

See Fig. 2 for key to systems.

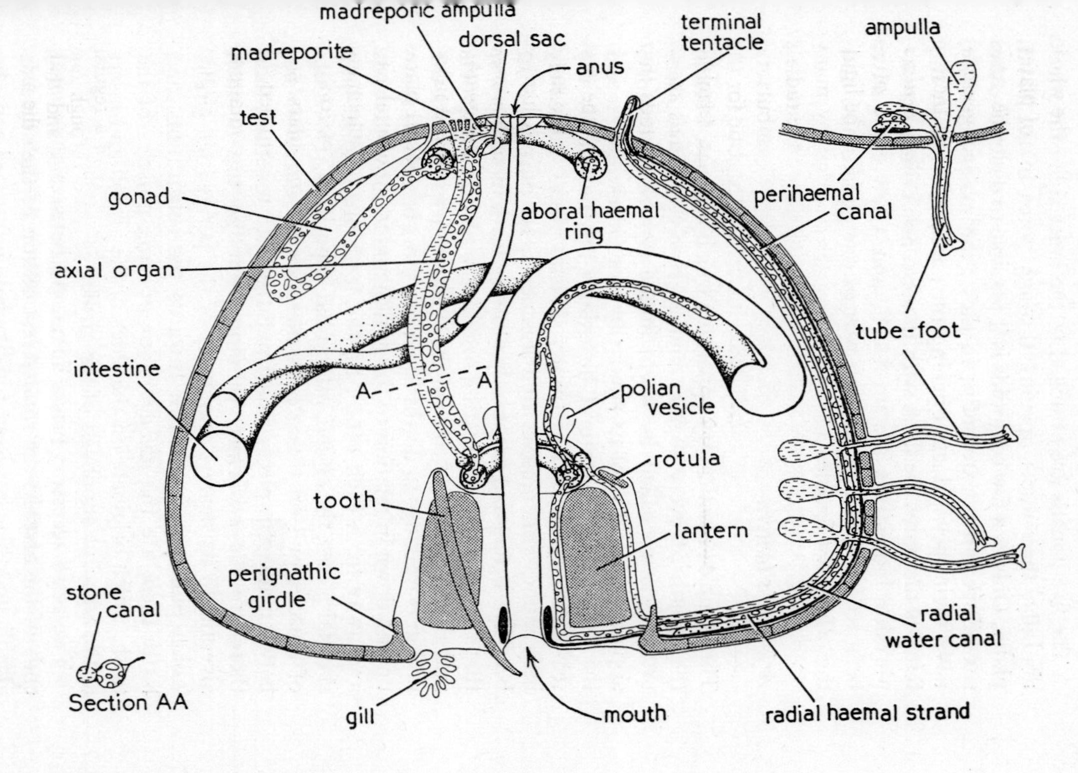
madreporic ampulla
dorsal sac
madreporite
anus
terminal tentacle
ampulla
test
gonad
perihaemal canal
aboral haemal ring
axial organ
tube-foot
intestine
A
A
polian vesicle
rotula
tooth
lantern
perignathic girdle
stone canal
Section AA
gill
mouth
radial water canal
radial haemal strand

The five gonads are attached to the inner side of the test aborally, their ducts opening through pores in the basal plates. Only in a few echinoids is it possible to determine the sex of specimens externally; in these, the males generally have their sperm ducts opening on papillae, while in the females the pores are flush with the test. Sometimes hermaphrodite individuals are produced, and a few cases have been recorded of urchins with part of one gonad male and the other part female.

Aristotle's lantern

The complex masticatory apparatus which is situated inside the mouth of every regular and every clypeasteroid sea-urchin was so called by Klein in the eighteenth century, apparently because of its similarity to a Greek lantern and the famous Greek natural historian's fascination for the sea-urchin. It consists of twenty skeletal pieces, intricately interbound with muscles and ligaments. Its purpose is to bear five strong and constantly growing teeth in such a way that they can rasp encrusting organisms, such as ectoprocts and algae, from the surface of the substratum over which the urchin is walking; to do this the teeth must be able to move up and down in relation to the mouth and to move towards and away from each other. So the teeth are held in long skeletal pieces, the *alveoli*, interradial in position. Three sets of muscles pull on these: first, there are the *interalveolars*, between adjacent pieces which pull the teeth together; then there are the *adductors*, running from the top of each alveolus to its nearest apophysis of the perignathic girdle, which pull the teeth down towards the substratum; and lastly there are the *abductors*, or opening muscles of the teeth, which originate on the sides of the alveoli and insert at the adjacent apophysis of the girdle.

Two small accessory pieces lie radially between the aboral ends of the alveoli: the *rotulae* rest on top of the adjacent alveoli, and the *compasses*, rod-like with bifurcated ends, lie on top of them. Each of these last-named pieces has a muscle near its inner end, joining it to its neighbour, and two ligaments at its outer end, attaching it to the two apophyses to either side of it. The set of compasses, muscles and ligaments serve a function only indirectly concerned with feeding: they act to pull the peri-oesophageal coelomic membrane up and down so that coelomic fluid is drawn into and out of those gills or respiratory structures which supply the muscles of the lantern with oxygen. Clypeasteroid urchins have no gills, and their peripharyngeal region is not cut off from the rest of the coelom; and in consequence this set of structures is missing.

The alimentary canal

Arising from the largest plates in the peristome are ten buccal tube-feet. These are almost entirely sensory in function, and can be seen exploring the substratum over which the urchin is moving. If it passes over encrusting organisms, the teeth are set working to rasp off the food. This then passes into an *oesophagus*, which ascends up the centre of the lantern and on to its aboral surface, where there is considerable enlargement into an *intestine*, running right round the inside of the body to its starting-point, doubling back on itself and running halfway round again, then ascending to the anus. A curious structure associated with the gut, the *siphon*, is present in echinoids, but does not seem to be found in other echinoderms. This is a tube which comes off the gut at the start of the intestine, runs along parallel to it until the point of doubling, then re-enters it. Its function is not certain; there is said to be a current in it, so it may be for removing water during digestion. There is an interesting developmental feature of the mesenteries holding the main folds of the intestine: they are horizontal, but start life as the vertical plate of tissue between the two posterior coelomic sacs of the embryo (p.120). The urchin's adult mouth is formed on one side of the larva, so this plate of tissue becomes horizontal in respect of adult orientation.

The coelom

Students who crack open an echinoid test are usually disappointed to find that most of the inside is a fluid-filled cavity, with the gut hugging the inside wall of the test and the axial complex hanging from the apical disk in the middle like a piece of black thread. Certainly, the perivisceral coelom is more extensive than in any other class. It is also considerably subdivided, there being two distinct cavities round the anus, a *perianal* coelom, and outside that a *periproctal* coelom. Then, surrounding the whole of the lantern, as already mentioned, is the *peri-oesophageal* coelom.

The tubular coelomic systems

The circum-oral ring vessels of these systems lie just above the lantern (Fig. 8), and the radial water vessels pass under the rotulae, down the sides of the lantern adjacent to the alveolar muscles and under the arches of the auricles before passing up the inside mid-line of the ambulacra, while the radial haemal strands pass down the *inside* of the alveoli of the lantern, then between adjacent alveoli before passing under the arches of the auricles. The various radial structures bear the same relationship to each other in the ambulacra as they do in the ophiuroids; that is, the epineural sinus lies between the plates and the radial nerve cord, the perihaemal (= hyponeural) sinus comes next, then the haemal strand and lastly, most internal, the radial water vessel. The oral haemal ring gives off a branch leading to the very extensive lacunar system in the gut walls, and there is an aboral haemal ring, surrounded by a perihaemal space, giving off branches to the gonads. One curious thing about the perihaemal system is that the circum-oral ring does not appear to be in communication with the radial perihaemal canals. An axial sinus is absent[36].

In echinoids the axial part of the haemal system seems to

be distinct from the axial organ. The organ terminates aborally in a head-piece, as in asteroids and ophiuroids, and this is, as in them, contained in a dorsal sac, probably the remains of a right axocoel.

The nervous system

The main nerve ring is situated round the oesophagus inside the alveoli of the lantern, and the radial nerves leave it just below the radial haemal strands, following their path between the alveoli, under the auricles and out to the ambulacra. Because these main nerve tracts are beneath the skeleton, they are somewhat isolated from the main sensory sub-epithelial plexus which covers the whole body, being joined to it only by tracts passing out of the ambulacral pores.

Apart from the general sensitivity of the sub-epithelial plexus, there are few organs of special sense. The balance organs, *sphaeridia*, are dealt with in Chapter 8. Only in the family Diadematidae, apparently, are there any special light-sensitive areas.

The evolution and adaptive radiation of the echinoids

There is no clear-cut evidence as to the origin of the echinoids. Time-relations do not allow their derivation from the Ophiocistioidea of the Silurian (p.160), the only other echinoderm group with any sign of a masticatory apparatus, since the first echinoid, *Bothriocidaris* (Fig. 9*a*), occurs in the Ordovician. Mortensen, the greatest authority on the echinoids, whose *Monograph of the Echinoidea*[44] is a monument to a life's study of the group, thinks that the echinoids arose from close to the *Stromatocystis*-like edrioasteroids (Fig. 22*c*) of the Cambrian, on the grounds of similarity of plate arrangement. As in the asteroids, such a derivation would mean principally an inversion, so that the mouth came to face downwards. It is not easy to visualize how this rather drastic change of habit might have occurred—it is

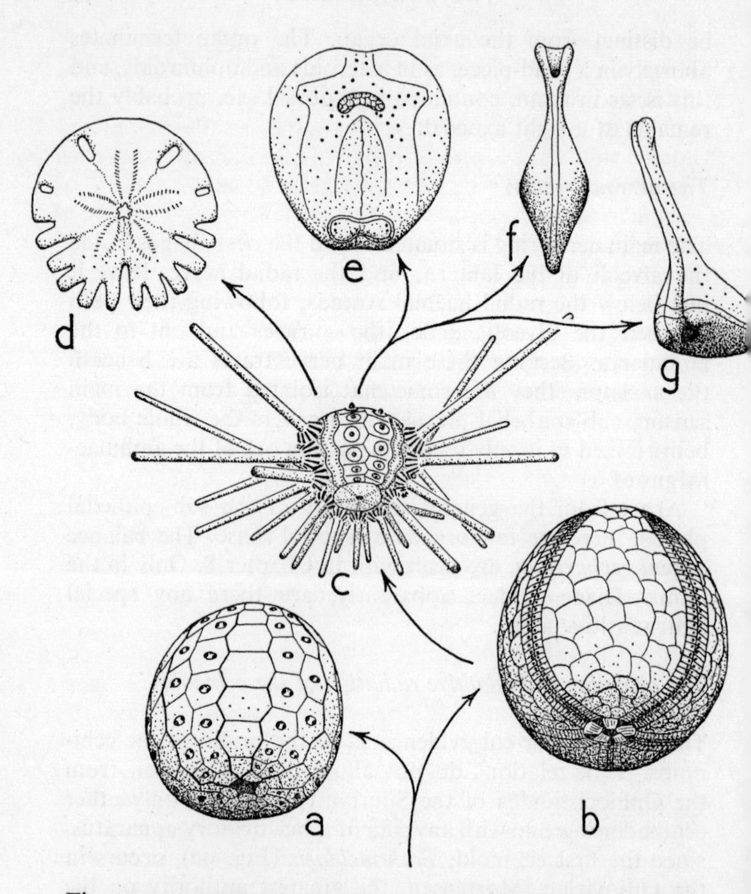

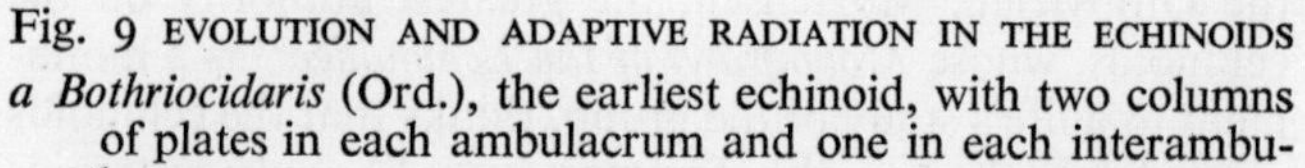

Fig. 9 EVOLUTION AND ADAPTIVE RADIATION IN THE ECHINOIDS

a Bothriocidaris (Ord.), the earliest echinoid, with two columns of plates in each ambulacrum and one in each interambulacrum.

b Aulechinus (slightly later Ord.), regarded as the echinoid closest to the ancestral type. Many plates in each interambulacrum.

c Cidaris, a typical regular echinoid, with two ambulacral and two interambulacral columns. Spines of the nearest interambulacrum removed.

hard to see it as anything but a sudden step, which would mean a correspondingly sudden change in feeding habits. Admittedly, if the edrioasteroid stock was ciliary-feeding (which would seem to have been the case) this method could still have been effective if the animal inverted, particularly if selection acted to extend the ambulacra on to the aboral (now upper) surface, as indeed did happen in some of the later edrioasteroids. Then, advantage would probably be gained by converting some of the peristomial plates into a dental apparatus for scraping organisms from the substratum.

So it is not surprising that even in the oldest known echinoids we see a primitive masticatory apparatus, forerunner of the complex Aristotle's lantern. Neither is it entirely unexpected that the tests of the earliest echinoids show signs of having been flexible, as were some edrioasteroids.

Though *Bothriocidaris* is the oldest known echinoid, it is not nowadays regarded as ancestral[37] because there are great differences between it and the early members of the next group to appear, the ECHINOCYSTITOIDA, from very slightly later in the Ordovician. The test of *Bothriocidaris* was probably fairly rigid, with two columns in each ambulacral area but only one column in each interambulacral; a surprising feature is that despite its age the recognizable imprints of tube-feet have been found associated with the pore-pairs of some ambulacral plates. In the peristomial region are five plates, believed to be the forerunners of teeth. In contrast, early members of the slightly later echinocystitoids, such as *Aulechinus* (Fig. 9*b*) and *Palaechinus*, usually had a slightly flexible test formed of two columns of

---

d to *g*, examples of recent and fossil irregular echinoids, spines removed.
d *Rotula*, a clypeasteroid sand-dollar, aboral view, showing lunules.
e *Spatangus*, a typical spatangoid heart-urchin, oral view.
f *Echinosigra*, a pourtalesiid spatangoid, oral view.
g *Hagenowia*, a spatangoid from the Cretaceous, in which a rostrum is present.

ambulacral and an indefinite number of interambulacral plates. From this, we infer that the first echinoid also had two columns of ambulacrals and probably an indefinite number of interambulacrals, which would fit in nicely with the theory of edrioasteroid ancestry. There is another feature in forms such as *Aulechinus* which suggests an edrioasteroid ancestry: they have an open ambulacrum, the radial water canal lying in a groove on the outside of the ambulacral plates, not in the interior of the test. The closure of the ambulacrum occurred during the evolution of the echinocystitoids.

After the establishment of the primitive echinoid there was, on the one hand, early specialization to the somewhat inflexible bothriocidaroid condition, which left no descendants, and, on the other, a time of experimentation from the Silurian to the top of the Permian in which various patterns of plate arrangement and methods of strengthening the test were tried. So we see a trend in some forms, such as *Melonechinus* (Carboniferous), towards a large number of ambulacral columns and in others, such as *Lepidesthes* (Carboniferous), to a large number of interambulacra. In none of these early forms is there a perignathic girdle for a strong attachment of the masticatory apparatus.

Some time in the Silurian one group of echinoids, the CIDAROIDA (Fig. 9*c*), restricted the number of columns in each ambulacrum and interambulacrum to two, and the group kept this pattern steadily for the rest of the Palaeozoic; they also evolved a perignathic girdle. Then, at the end of the Permian, something remarkable happened to the environment, something which probably happened the world over and which had the effect of apparently wiping out every echinoid genus except one, belonging to the cidaroids, called *Miocidaris*. Why this was the only one to survive nobody can say with any certainty, but it alone entered the Mesozoic era to provide the stock from which all subsequent echinoids arose. By Upper Triassic times regular echinoids other than cidaroids were also appearing, and there seems to have been a rapid deployment in the Jurassic, culminating at the present day in some sixteen families[37]. Most of these live on rocky bottoms, feeding on surface detritus and organisms which can be chiselled off by the teeth, and protecting themselves by means of spines and pedicellariae, many of which are poisonous. Several genera of various families, however, such as *Paracentrotus* of the Echinidae, may excavate cavities for themselves in the rock, in order to gain protection from wave-action. They do this by clinging to the rock-surface by their tube-feet and continuously abrading the rock with their spines and teeth[47]. They feed either on particles which are washed to the bottom of the cavity, or they may turn in the cavity to bring the mouth uppermost. The interesting point about these boring species is that some populations of apparently the same species do not bore, even though a similar type of rock may be available to them. For instance, *Paracentrotus lividus* on the coast of Ireland and North-west France may honeycomb the rock with its cavities, while members of the same species in the Mediterranean hardly ever bore into the rock.

At about the beginning of the Jurassic we see the first signs of a break in the almost perfect radial symmetry of the urchins, a step which gave them definite front and back ends and which opened up new habitats hitherto forbidden to the group. There appear to have been two main attempts at irregularity: one, grouped as the GNATHOSTOMATA, retained the Aristotle's lantern and girdle, whilst the other, the ATELOSTOMATA, dispensed with it. The earliest irregular urchin to appear in the record is the atelostome *Pygomalus*, from the Sinemurian stage of the Lower Jurassic; the earliest gnathostomes were *Holectypus* and *Galeropygus* from the slightly later Domerian. Both gnathostomes and atelostomes persist to the present day, the gnathostomes containing two orders, HOLECTYPOIDA and CLYPEASTEROIDA, and the atelostomes being represented mainly by the SPATANGOIDA (Fig. 9*e*, *f*, *g*). Though there can be little doubt of their separate derivation, the two groups of irregulars show many parallel changes, the chief one being the marked specialization of tube-feet on various parts of the animal for particular jobs: in both, for instance, we see the evolution of special areas of non-extensile tube-feet on the dorsal surface for respiration only, and in some members of both we see the complete loss of suckered tube-feet. Both these trends are associated with the adoption of a burrowing habit, with its consequent protective advantages, a habit exploited to the full by the spatangoids. Among the gnathostomes the clypeasteroid sand-dollars achieve probably the greatest specialization, some, such as the Key-hole Urchin, *Rotula* (Fig. 9*d*), becoming remarkably flat and possessing holes through the test, from the upper surface to the lower, probably to aid burrowing. They sit in the substratum either flat and just below the surface, rather like a flatfish, or at an angle with an edge of the test protruding to face oncoming currents, presumably as a food-catching device. The tiny British urchin *Echinocyamus*, sometimes found inshore on shell-gravel or coarse sand but more often offshore, is a clypeasteroid. But it is not a sand-dollar; these flat forms do not occur in this country.

The spatangoids (Fig. 9*e*) are far more ingenious in their mode of burrowing. Some of them, such as the common British heart-urchin of sandy shores, *Echinocardium*, have tube-feet on different parts of the body specialized for burrow-building and feeding (p.113) and they can burrow at depths of anything up to five times their own height, all the time maintaining the vital connexion with the surface of the substratum for food and oxygenated water[45, 46]. The mouth in spatangoids is usually near the front of the underside, and the anus usually on the posterior surface. Feeding is by a combination of ciliary action and the activity of sticky tube-feet, which pass material from the substratum into the gut where the organisms adhering to it are digested off; the particles of the substratum are then voided from the anus. The most peculiar spatangoids are probably the deep-water pourtalesiids[42], such as *Echinosigra* (Fig. 9*f*), in which the mouth is at the very front and the anus almost at the rear, giving it a strong resemblance to a holothuroid in shape. Another curious form was the Cretaceous *Hagenowia* (Fig. 9*g*) in which the aboral surface is drawn out into a rostrum.

6

THE HOLOTHUROIDEA

THE sea-cucumbers are eleutherozoan echinoderms that lie on one side with the mouth and anus at the opposite ends of a cylindrical body. By no stretch of the imagination can most of them be described as elegant: they are limp and slimy to the touch, and some exude special sticky strings to entangle anything (and anyone) who meddles with them. Their skins are usually tough and leathery, with embedded spicules. Yet the Chinese relish a meal of *trepang*, which consists of sun-dried body walls of several species of *Holothuria*, *Stichopus* and *Thelenota*. 'It is employed,' says Forbes[5], (' . . . in) the preparation of nutritious soups, in common with an esculent sea-weed, sharks' fins, edible bird's-nests, . . . affording much jelly.' Certainly, trepang appears to be highly nutritious, and, as its protein constituents are almost completely broken down by pepsin, highly digestible.

Holothuroids feed by means of the buccal tube-feet. Either they move slowly over the sea-bottom, sweeping the area ahead or under them with their sticky arborescent tentacles, or they nestle in nooks in the rock, and sweep the 'forecourt' for settling organisms, or they burrow into the soft substratum and sit with their buccal tube-feet forming a collecting funnel for the rain of detritus. A few holothuroids are pelagic, e.g. the elasipod *Pelagothuria* (Fig. 11*g*), in which parts of the body wall and tube-feet are extended to form a swimming membrane; they apparently feed on plankton. These forms are much more beautiful than the others, and closely parallel the jellyfishes in morphology and ecology.

General body plan

The ambulacra pass down the body lengthwise (Fig. 10). They are arranged so that three ambulacra lie ventrally (the *trivium*) and two dorsally (the *bivium*). When tube-feet are present along the body, that is, in all groups except the Apoda, those of the trivium are generally suckered, while those of the bivium are papillate. Often, as in *Holothuria*, the tube-feet arise all over the body, not restricted to ambulacra, so that the trivium forms a creeping sole and the bivium a dorsal sheet of mainly sensory papillae. Only one gonad is present, in the mid-dorsal interambulacrum, opening by a gonopore just behind the mouth. In those in which there is a madreporite this too is anterior.

It is at the posterior end of the alimentary canal that the main differences between these and other echinoderms are found: here, special respiratory structures and, in some, the special defensive organs are present as outgrowths of the cloaca, to be explained later. They occupy a large part of the general body cavity.

On the outside of the body there are no spines, no cilia and no pedicellariae. The body-wall itself contains the usual calcareous elements but in this case they are severely reduced, to spicules in the shape of rods, crosses, hooks, anchors or wheels. Sometimes they form a protective armour, like roof tiles. The shape and combination of these spicules are used extensively by taxonomists. It would be interesting to know how each shape fits into the pattern of

---

Fig. 10 BASIC ANATOMY OF A HOLOTHUROID

Diagrammatic vertical section through the body of a generalized holothuroid, based on *Holothuria*. The section is taken through the ventral ambulacrum, of which three tube-feet and their ampullae are shown. In the posterior region only one of the pair of respiratory trees and cuvierian organs is shown.

See Fig. 2 for key to systems.

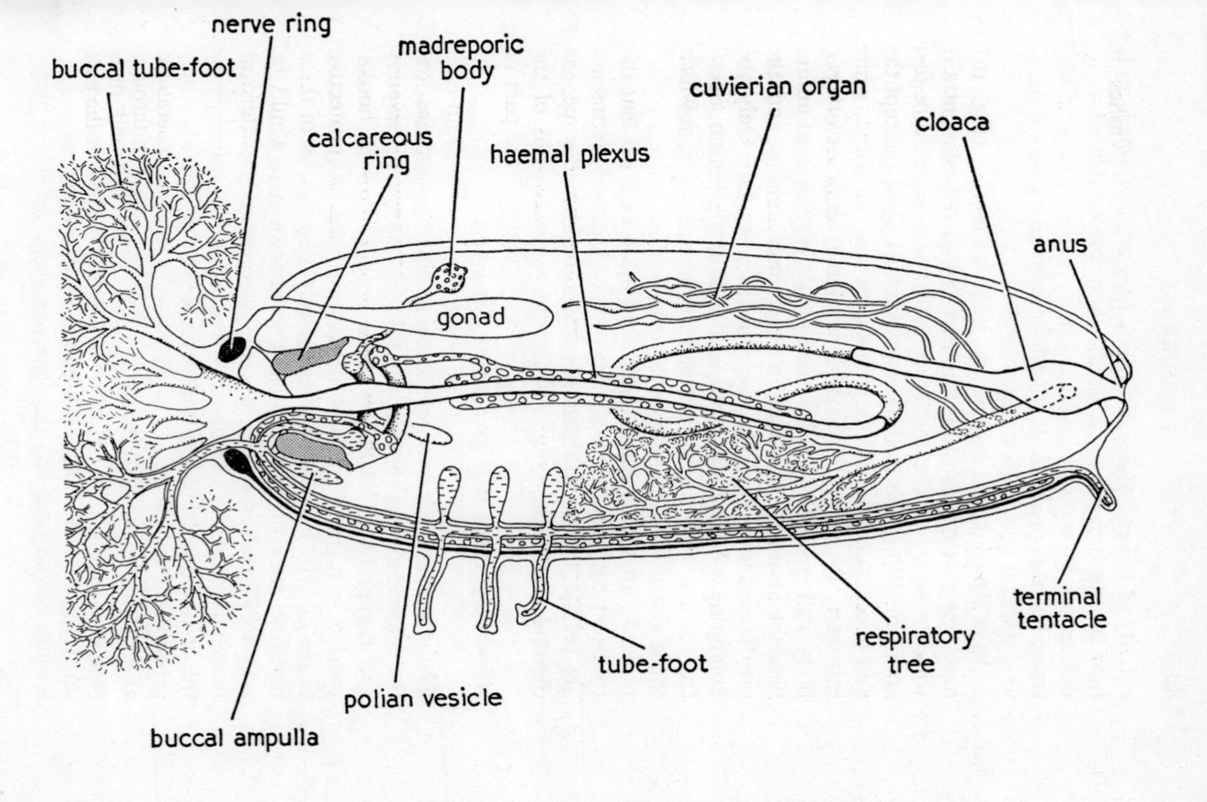
nerve ring
buccal tube-foot
madreporic body
calcareous ring
haemal plexus
cuvierian organ
cloaca
anus
gonad
terminal tentacle
respiratory tree
tube-foot
polian vesicle
buccal ampulla

activity of the soft tissues in which they are contained, and how specific differences in spicule shape are correlated with differences in ecology and behaviour between close relatives; but this problem of functional taxonomy would be a very difficult and challenging one to tackle.

As in asteroids and ophiuroids, there is a complete ring of much larger ossicles round the mouth and oesophagus, to which muscles of the buccal tube-feet and body wall are attached. When the feeding animal is disturbed the mouth and buccal tube-feet can be withdrawn into the body on an introvert, by contraction of longitudinal muscles of the body wall inserted at the calcareous ring. Extension is brought about by contraction of circular muscles, acting on the coelomic fluid. As one would expect from animals inhabiting a variety of ecological niches, the buccal tube-feet show a wide range of form. To some extent this follows the division of the holothuroid orders, classified on other grounds. In the Dendrochirota and Elasipoda the buccal tube-feet are mostly dendritic; in the Aspidochirota they are peltate, in the Molpadonia digitate and in the Apoda digitate or pinnate.

Alimentary canal and its accessory structures

The mouth leads into an oesophagus which passes through the calcareous ring to a slightly enlarged part, the stomach, and thence to an intestine whose loops follow the same pattern in relation to the orientation of the ambulacra as those of crinoids, though drawn out longitudinally. The intestine ends in a cloaca, and it is from here that the accessory structures arise, when they are present. A pair of these structures, the *respiratory trees* (Fig. 10), arises from the cloaca in all holothuroids except the Apoda. They are arborescent tubules, lined with the same layers as the gut and ending in thin-walled vesicles. The cloaca, pumping at a rate of six to ten pulses per minute, and working in conjunction with an anal sphincter, forces water into the lumina of the trees, so that gaseous exchange takes place between the sea-water and the coelomic fluid. A curious example of animal relationship occurs here: there is a teleost fish, *Fierasfer*, which lives in the main stems of the trees of aspidochirotes. Its head protrudes from the anus and it uses its host apparently solely for shelter, catching its own crustacean food during night sorties.

In aspidochirotes there are two groups of special defensive *cuvierian organs* branching from the bases of the respiratory trees (Fig. 10). These are long structures, sometimes hollow, with an outer layer of special cells, two layers of muscle fibres just below the surface, and a thick layer of collagenous connective tissue. When the animal is irritated the structures are extruded steadily from the anus and acquire a sticky surface, thus trapping the aggressor. The British species *Holothuria forskäli* will perform in this way when handled, and the mess one individual can make must be seen to be believed. The process has given rise to the common name 'cotton-spinner' for *Holothuria*. The mechanism of extrusion is believed[52] to have two components: first, water from the lumen of the respiratory trees is forced into the tubules, the proximal ends of the tubules are shut off by muscles, and the muscles in the walls of the tubules contract, elongating them by hydrostatic pressure. The cloacal wall then splits at a weak point formed by the cross-over of circular and longitudinal muscles, so that the tubules are forced out through the anus to elongate steadily into threads sometimes several feet long. The elongation splits the covering cells, freeing a sticky substance, and the collagen fibres keep the threads from breaking, thus making a formidable trap for the unwary. The process has its parallel in the holothuroids without cuvierian organs: these forms when annoyed split the intestinal wall near the anus but simply spew out the gut and respiratory trees, to regenerate them again later. The significance of this operation in the life of the animal is not clear, but possibly the gut and associated organs provide a tasty meal for a predator bent on having one, leaving the tougher remains of the animal to creep quietly away.

The coelom

This is subdivided by mesenteries into four cavities: the main perivisceral part, an anterior perioesophageal part with a peribuccal part within it, and a posterior perianal part. The gut is held in the coelom by mesenteries, and in the Apoda these contain special *ciliated urns*, which apparently collect waste matter enclosed in coelomocytes. What happens to the waste subsequently is not clear, but some authorities say it is passed out through the body wall.

The tubular coelomic systems

Of these, only the water vascular and haemal systems are represented to any extent, though a channel corresponding in position to the perihaemal canal has been described in the ambulacra of some holothuroids. There is normally no axial complex as such, and though the stone canal is always present, it very often does not open at an exterior madreporite but terminates within the coelom at a *madreporic body*, a swelling with many ciliated pores piercing it. The water ring lies posterior to the ring of ossicles (Fig. 10) and gives off a number of polian vesicles (one in *Holothuria forskäli*, up to thirty in some others) as well as the five radial canals. The canals pass forwards, down the sides of the mesentery forming the wall of the perioesophageal coelom, give off the lumina of the buccal tube-feet and their ampullae and continue inside the body to the posterior end, where they terminate as the lumina of the terminal tentacles. The class Apoda is an exception to this: in it there are no radial coelomic vessels at all, only the circum-oral ring vessels and buccal tube-feet.

The haemal system is well developed, mostly in association with the gut, where there are two main branches on either side of it joined by a mass of capillaries, a *rete mirabile* ('wondrous network'), over the surface of the gut wall. The oral haemal ring, aboral to the water vascular ring, gives off the usual five radial haemal strands, lying with the same relationship to the radial elements as in other echinoderms. The axial organ is probably absent, though some say it is represented by a small knot of tissue arising from the haemal ring adjacent to the origin of the stone canal.

The nervous system

The main nerve ring, as in other echinoderms, lies nearer the mouth than do the coelomic ring components, and is embedded in the peristomial membrane close to the bases of the buccal tube-feet. It sends radial branches out under the radial components of the calcareous ring, and these branches ramify into the general sub-epithelial plexus of the body wall, continuing into the tube-feet. The system just described is the *ectoneural system.* But the radial nerve cords are not single components: a section through the ambulacrum of a holothuroid shows that in addition to the ectoneural part there are aggregations of neurones internal to it and to some extent ramifying with it. This is the *hyponeural system*, homologous with that of asteroids (Lange's centres), and possibly also with the deep oral system of crinoids. It is mainly motor, but it too receives fibres from the sub-epithelial plexus of the body wall.

The main sense organs of holothuroids are the tactile and chemosensory cells scattered in the epithelium, and, in some Apoda, aggregations of nerve cells in special warts on the body surface. In some Apoda, molpadids and elasipods there are, in addition, statocysts embedded in the body wall close to the point where the radial nerves leave the nerve ring. These are tiny fluid-filled spheres containing calcareous statoliths, the differential movement of which is registered in special nerves and conveyed to the radial nerve, presumably to right the animal if its orientation is upset by external forces.

The ambulacrum

From what has been said about the radial components of the various systems in the ambulacra it will be evident that holothuroids, in common with ophiuroids and echinoids, possess a *closed ambulacrum*. Like them, the enclosure of the radial elements inside the body wall can be followed during development, and results in a small canal, the epineural sinus, remaining between the body wall and the radial nerve. We have seen that in each of the classes with a closed ambulacrum this sinus is present, as is an equivalent one external to the nerve ring round the mouth. Whether this sinus communicates with the coelomic spaces is not known, but one must conclude that for physiological reasons the nerve cords must be bathed on either side by fluid, on the inside by that in the radial perihaemal sinus and on the outside either by the sea-water (open system) or by the fluid in the epineural sinus (closed system).

Evolution and adaptive radiation of the holothuroids

Because their skeletal plates are reduced to mere ossicles, we have little or no picture from the fossil record of holothuroid origin and evolution. All we can say with certainty is that their spicules first appear in the Carboniferous[50], but as to their descent we must rely solely on development and anatomical features, despite the shortcomings of such evidence.

The circum-oesophageal calcareous ring, consisting of five radial alternating with five interradial ossicles, could have been inherited from the stock which also gave rise to the stelleroids and echinoids, both of which possess similar structures. But nobody can say whether the holothuroids passed through a stage in which the mouth was directed downwards: circumstantial evidence suggests that they did not. The presence of a single gonad, with its pore and the hydropore in the same interradius, is regarded[7] as primitive, that is, as persisting from the early pelmatozoan condition, and this link is further seen during their development: in the larvae of both holothuroids and crinoids a vestibule is present into which the primary tube-feet protrude during their formation (pp.124 and 128). So, tentatively, we may say that the holothuroids possibly arose from the eleutherozoan stock at an early stage, soon after it had diverged from the crinoids; they may have arisen from free edrioasteroids, like *Stromatocystis* (Fig. 22*c*), or they may represent an independent achievement of the eleutherozoan condition, the available evidence being insufficient to say further.

The holothuroids show a surprising degree of adaptive radiation. Some members are benthic, some pelagic and some burrowing; they are found in littoral waters or at great depths; some are exceedingly sluggish, while others exhibit a fair turn of speed when necessary.

The benthic, creeping holothuroids are mainly but not entirely included in the orders DENDROCHIROTA and ASPIDOCHIROTA. The dendrochirote *Psolus* (Fig. 11*a*) has its ventral trivium flattened to form a creeping, muscular sole for locomotion, with very few tube-feet involved, rather like the foot of a mollusc, while the aspidochirote *Holothuria* has tube-feet over the whole ventral surface, and these are the locomotory organs. The mouth in these creeping forms may be at the anterior end, so that the oral tube-feet can sweep the sea-floor for food, as in some species of *Cucumaria* (Dendrochirota) and *Holothuria* (Aspidochirota), or it may point dorsally, as in *Psolus*, probably to collect falling detritus. Members of the other order with tube-feet present down the length of its body, the ELASIPODA, show a variety of structures forming sails and other flotation devices: these forms are mainly bathypelagic, and are known mainly from the collections of major expeditions[55]. One of the most curious is *Pelagothuria* (Fig. 11*g*), known principally from collections made by the *Albatross* and other vessels, in which parts of the body wall are extended as radially arranged papillae supporting a web[48]. The mouth, surrounded by oral tube-feet, is directed upwards and the

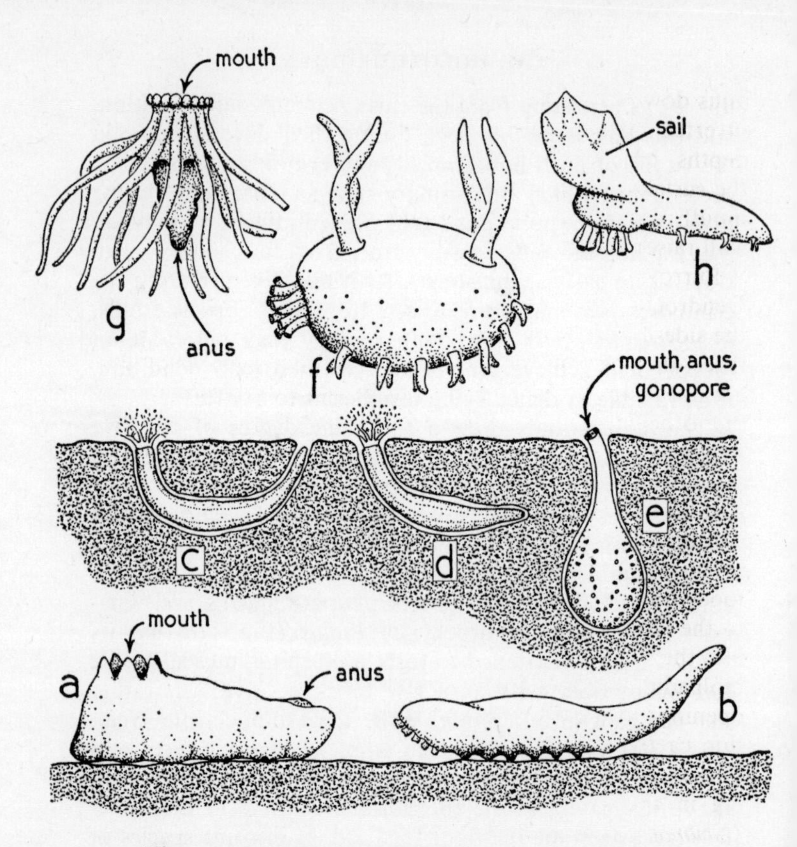

Fig. 11 ADAPTIVE RADIATION IN THE HOLOTHUROIDS

a *Psolus*, a dendrochirote, with mouth and anus directed upwards.

b *Psychropotes*, an elasipod, with mouth and anus directed downwards.

c to *e*, some burrowing holothuroids.

c *Cucumaria*, a dendrochirote which uses anal respiration.

d *Synapta*, an apodan, which lacks respiratory trees, and does not appear to need a U-shaped burrow.

e *Rhopalodina*, an unusual dendrochirote, in which mouth, anus and gonopore all open at the end of a spout.

f *Scotoplanes*, a dendrochirote with large dorsal papillae.

g *Pelagothuria*, a pelagic elasipod.

h *Peniagone*, an elasipod with part of the body wall raised as a sail.

anus downwards, so that the form somewhat resembles an inverted medusa. Some species have been taken at great depths, others at moderate depths and others actually at the surface. Another form is *Peniagone* (Fig. 11*h*), with the mouth directed downwards and part of the dorsal body wall raised into a sail.

Burrowing holothuroids are found mainly in the order Dendrochirota and the two orders lacking tube-feet down the sides of the body, the MOLPADONIA and APODA. Of these three, the latter alone lacks respiratory trees, so its members are not restricted to U-shaped burrows, as are members of the other two. The dendrochirote *Thyone* burrows by wriggling the middle of the body, so that this part alone sinks below the surface, while *Cucumaria*, in the same order, goes down head-first until it is buried, then moves round and upwards to break surface a few inches in front of the original hole. It then extends its oral tube-feet to catch food. If the original hole falls in, the tail will be pushed up to the surface again (Fig. 11*c*), so that water can be taken into the respiratory trees through the cloaca. The burrowing molpadids, such as *Molpadia* and *Caudina*, also require two openings to the burrow, and these forms usually possess a thin caudal region which can be pushed to the surface of the substratum for respiration and defaecation. It is interesting in this connexion that in the British dendrochirote *Cucumaria elongata* the faeces are voided with some force, so that they do not drop back into the burrow to foul the water used for respiration. The apodans, such as *Synapta*, do not possess respiratory trees, and though they too push themselves into the substratum head-first and move round so that their head protrudes from the surface, they do not need to keep the anus in contact with the water (Fig. 11*d*): respiration in this case is probably solely by diffusion across the walls of the oral tube-feet.

7

PENTAMEROUS SYMMETRY AND THE ECHINODERM SKELETON

The origin of pentamerous symmetry in the echinoderms

The main character separating the echinoderms from other phyla is the presence in all but a few non-crinoid Pelmatozoa of pentamerous symmetry, though some forms (holothuroids and irregular echinoids) have bilaterality superimposed. Why this relatively constant feature should have arisen has never been adequately explained; D'Arcy Thomson, in his classic work *On Growth and Form*, quotes the mathematics of the arrangement, but makes no attempt to explain its origin; Breder[56] also treats the shape geometrically only.

Even though the echinoderms are one of the few phyla with the sort of skeleton that fossilizes easily, their fossil history gives us little information useful in helping to decide when pentamerism first arose. The first echinoderms to appear in the fossil record (the Eocrinoidea from the Lower Cambrian) were already pentamerous, although the Heterostelea, which first appear as fossils later, were not. They may, of course, have secondarily assumed bilateral symmetry; or they may not be echinoderms at all, though this seems unlikely. Later in the time-scale the Cystoidea (Ordovician to Silurian) appear, and in this group we can trace what appears to be a second attainment of pentamery in the thecal plates. At first they show the typical pelmatozoan condition of radially symmetrical theca with the mouth in the centre of the upwardly directed surface and the anus on the side of the theca but with a haphazard arrangement of thecal plates (e.g. *Aristocystites*, Fig. 19*b*, and *Echinosphaerites*, Fig. 19*g*). Later, the theca tends to have fewer plates arranged mostly in cycles of five (e.g. *Lepadocrinus*, Fig. 19*i*). The ambulacra, too, show a tendency to assume pentamerism: there were apparently only three food-collecting brachioles in the early cystoids such as *Echinosphaerites*, in which the brachioles arose from close to the rim of the mouth, but in those in which the food grooves extended over the theca and the brachioles arose from facets along the sides of the food grooves and further from the mouth (e.g. *Glyptosphaerites*) the number of grooves appears to have settled at five. But there is one very curious thing about the thecae of those animals, such as *Aristocystites* and *Echinosphaerites*, in which the main body plates were irregular and non-pentamerous: the anal pyramid was invariably made up of five small plates, and this was the only part of the animal which showed any trace of pentamerism, as far as we can see.

The development of a calcareous theca is obviously to confer protection on the settled echinoderm which cannot move at all, or only very slowly, to escape predators. But before the theca has become strong and efficient as a protection, that is, just after metamorphosis when the plates are not properly bound together, a nudge from a passing animal may be sufficient to dislodge them. It seems to be the case that the chief planes of weakness in any plated theca lie along the inter-plate sutures, so it is desirable to have as few sutures as possible, and these as short as possible, and it appears likely, from all the evidence available, that these two mechanical demands may have influenced the development of pentamerism in echinoderms. That plates do become dislodged occasionally during early development is obvious from the recent and fossil monstrosities which turn up from time to time: in most cases it seems to be the sutures which have given way. In these cases, only if the damage is fairly slight and does not completely upset the subsequent plate growth can the animal survive to maturity.

The tests of all living classes of thecate echinoderms have a basically similar mode of development in their apical (aboral) regions. A central plate, usually with the anus opening through it, is laid down and surrounding it is a ring of five plates, the basals of crinoids and the genitals of asteroids, echinoids and ophiuroids. Gordon has shown that during the development of the regular sea-urchins *Psammechinus miliaris* and *Echinarachnius parma* the first six plates to form during metamorphosis (the anal plate and the ring of five genitals surrounding it) occupy nearly the whole of the aboral surface, and hence this shield will be required to take a large part in protecting the young urchin from the rigours of the sea-bottom. After the first ring of plates, which are inter-radial in position, a second ring, the radials, is laid down outside and alternates with them; from this primordium, the other plates of the theca are budded off in various ways, and it follows that if the theca is to be radially symmetrical each ring will have to be composed of plates of identical shape and size. Assuming that the plates must cover the theca completely with no gaping holes between them, one can put forward a strong case for the use of five plates in the first ring. Suppose the number of plates in the first ring were three, then each plate would be roughly of the shape shown in Fig. 12*a*, and there would be three main lines of weakness, such as the one shown by the arrows. The angles between the continuations of these lines (ideally 150°) are so obtuse that for practical purposes the lines may be considered nearly straight. In other words, though the sutures are few, they are long. If the first ring consisted of four plates (Fig. 12*b*), there would be two main lines of weakness across the apical region, one of which is indicated in the figure; although both these lines are interrupted by the central plate, the two interplate sutures of each are in a straight line, and if the anus opens through the central plate this will further add to the weakness of each line. Similarly, if the first ring were composed of six plates

(Fig. 12*d*), each perradial suture and the suture opposite to it would form a line of weakness. Also, with this arrangement the number of sutures is high, and this, clearly, should be avoided.

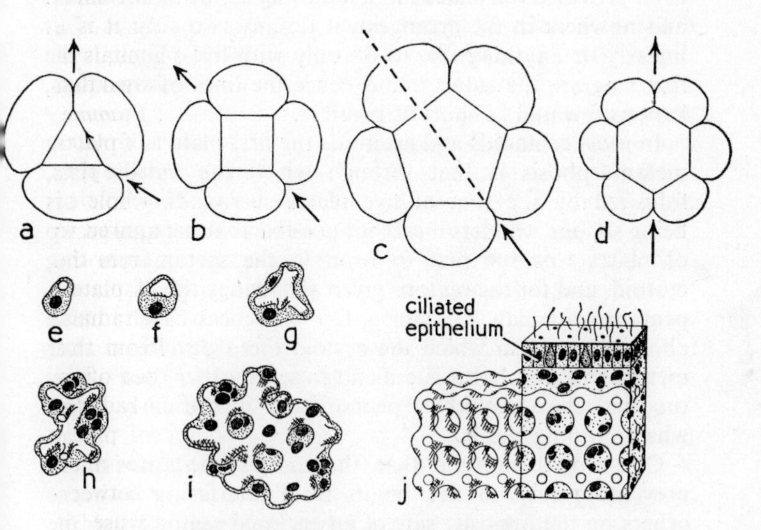

Fig. 12 THE ECHINODERM SKELETON

a to *d*, diagrams to show a suggested reason for the establishment of five plates in the first ring of the apical disk of an echinoderm.

a With three plates, and a line of weakness shown by the arrows.

b With four plates, opposing sutures in line.

c With five plates, the actual situation, in which no suture is in line with any other.

d With six plates, opposing sutures in line.

e to *j*, the formation and structure of the echinoderm skeleton.

e An ossicle-producing cell laying down the first grain of calcite as a single crystal.

f The crystal enlarges.

g The cell divides and the crystal grows.

h Further division and growth.

i Typical reticular pattern now recognizable.

j To show the relation of the skeleton, the tissue which forms it and the external epithelium.

But if the number of plates in the first ring is five (Fig. 12*c*), none of the sutures is in line with any other, even on the far side of the central plate; in fact, the continuation of each perradial suture will exactly bisect the plate opposite to it, provided the plates in the first ring are of uniform size, and nowhere in the arrangement do any two sutures abut linearly or anything like it. So only with five plates in the first ring are the sutures, and hence the lines of weakness, kept as few and as short as possible.

In most echinoids and asteroids the first plate to form on metamorphosis is that through which the anus opens, followed by the ring of five plates surrounding it. This being so, one wonders if it is not possible that the anal group of plates was the first to form in the metamorphosing cystoid, and for the reasons given above they too assumed a pentamerous plan. Of course, few inferences can be made about the way in which the cystoid theca developed after metamorphosis, but it is difficult to see another reason why the anal spire should be pentamerous in the early forms where nothing else is.

One would suppose that the use of pentamerism in preventing each of five points from interfering with the others on the opposite side of an enclosed region would be shown by other groups of the animal kingdom, and indeed this does seem to be so in the Priapulida, the only other phylum exhibiting any marked pentamerism. In *Priapulus caudatus*, for instance, the buccal teeth are arranged in concentric circles of five, and this very curious arrangement has never been adequately explained. On the argument put forward above, it seems likely that here also it would be a disadvantage for the teeth to be in rings of two, four or six, for with any of these arrangements each tooth would interfere with the one opposite when the animal retracts the pharynx to feed in the prey. With three or five teeth in each ring, however, this interference would not occur, and only with five teeth in each ring would the aperture be anything like circular. It is particularly interesting that in those polychaets which also use a toothed pharynx to feed in their prey, such as *Nereis*, the pharyngeal teeth are also in five main groups.

The echinoderm calcite skeleton

In addition to possessing an almost unique type of symmetry, the echinoderms have another feature in which they stand alone in the animal kingdom: the structure of their calcite skeleton. What is this curiously unique structure? Why should it be restricted to one phylum? And could it have anything to do with the adoption of five-rayed symmetry?

The answer to the first question is this: each skeletal element, be it a plate of calcite helping to make up the theca, a spine, or one valve of a pedicellaria (p. 100), is composed of a single crystal of calcite. Each crystal starts as a tiny grain within a skeleton-forming cell (Fig. 12*e*) and grows until it occupies nearly the whole cell except for the nucleus[57]. This then divides (Fig. 12*g*) and the daughter nuclei move apart, so that the cell, and the grain of calcite it contains, can grow. The division is repeated again and again, but the grain does not grow as a solid object; instead, it accresces in many directions, rather as the girder skeleton of a building is constructed, so that a crystal network is formed, the nuclei of the syncytial cell mass producing it lying in the spaces within and around it (Fig. 12*h*, *i*). In this way the crystal can be added to in any direction and can ultimately assume any shape that is needed. The reverse process can happen too: spicules can be eroded away in places if their shape needs to be altered during the growth of the animal. As an example of this, the plates round the peristome of most echinoids are altered in shape by resorption as the animal grows[39].

The reticular pattern possessed by the skeleton has three possible advantages: first, it will be relatively lighter and more economical in material than a solid crystal of the same dimension; secondly, the holes in it will provide an insertion for the connective tissue which binds the elements together;

and, thirdly, the holes may offer a resistance to a shearing or splitting force by providing an interruption in the cleavage plane of the crystal. If you snap a plate or spine from a recent echinoderm in two, it will seldom break across a single cleavage plane but will rather fracture along an irregular path, depending on the different thicknesses of the component struts; and the broken ends of the struts will show different cleavage angles. The situation is rather different, however, where a fossil spicule is concerned. Here the original calcite is replaced first, and then the spaces within the crystal, where the skeleton-forming cell mass originally lay, become filled with secondarily deposited calcite; this attaches itself to the calcite already there along the existing crystal plane, so that now the skeletal element becomes a more or less solid continuous crystal which may even be continued beyond the limits of the original element. In consequence, one would expect a fossilized echinoderm plate to fracture along the expected plane rather better than one from a recent form, and this is indeed the case. In addition, in a well-preserved fossil the original reticular pattern will still be visible, so that palaeontologists are justifiably confident of their ability to put echinoderm remains into their phylum with some certainty.

The answer to the second question, why only the echinoderms have this type of skeleton, does not seem to have been answered. The fact is that other animal groups which utilize calcite as a skeletal material either use solid spicules, as in the sponges, or incorporate extraneous material, such as sand, clay and other chemicals, into the skeleton, as in protozoa. The further question may be asked: is there any evidence that other groups using this skeletal system have been evolved independently in the past? And in answer one could point to such enigmatic groups as the Machaeridia, Cycloidea and Cyamoidea, the possible interpretation of which is discussed in Chapter 12, and say that here might be somewhat unsuccessful attempts to evolve a calcareous skeleton on the same lines.

The answer to the third question, whether there is a

possible connexion between the unique spicule structure and the unique five-rayed symmetry, is probably yes. The strength of the thecal plates composed of single crystals in a reticular shape is most likely such that the weakest places in the theca are the sutures between the plates; and it has already been shown in the previous section that this is one of the prerequisites of the most plausible explanations for pentamery.

8

SPINES AND PEDICELLARIAE

Spines

The echinoderm structures we shall consider in this chapter are those to which the phylum name refers (Greek 'spiny skinned'). Of the living forms the asteroids, ophiuroids and echinoids possess spines, while we can be sure that their ancestral group, the fossil edrioasteroids, also possessed them, because the tubercles which bore them are preserved on some specimens. It is not at all certain that the earliest echinoderms had spines: tubercles have not been found in cystoids, blastoids and heterosteles, and there are no spines on crinoids of today, and no evidence that they ever had any. The holothuroids must have secondarily lost them, if we regard the class as derived, like other Eleutherozoa, from the edrioasteroids.

The simplest spines are found in the asteroids. Most asteroids have spines on their adambulacral plates, many have them on the other lateral and also the aboral plates, but none has spines in the ambulacral areas. The spines (Fig. 13*a*, *b*) consist simply of a calcareous piece held in an erect position on the underlying plate by a number of muscles and covered by the general body epithelium, though this may get secondarily worn off distally. Gland cells secreting mainly mucus but sometimes also poison may be embedded in special spaces in the calcite material. A second type of spine is found chiefly in the Phanerozonia. In Chapter 3 it was mentioned that important currents pass from the aboral surface of most burrowing asteroids to the oral surface both for respiration and for feeding. This current is largely created by special ciliated spines, *clavulae*, in the interstices of the marginal plates on the arm-sides. Each clavula (Fig. 13*c*) is somewhat club-shaped, may have mucous glands distally and has two bands of cilia down its length, one on either side, so that currents are created in one direction only. Also in the phanerozone starfish are special groups of spines on the aboral surface, part of structures called *paxillae*. These structures consist of special raised body ossicles, each of which has a crown of spines arranged in such a way (Fig. 13*d*) that they can be raised vertically or lowered through a right angle to form a covering to the aboral surface. Gland cells in the spines exude a mucus which forms a sheet across the closed paxillae so that a cavity can be kept open around the burrowing animal for respiration, etc. Another feature is found in some phanerozones: some spines in forms such as *Pteraster* are held to their neighbours by a membrane (Fig. 13*e*) which presumably can be moved like a fan to create currents for feeding and respiration; then in *Hymenaster* the marginal spines, with fans, form membranes between the arms, possibly to increase the surface area for food collection. In the asteroids it may be difficult to tell the difference between spines and pedicellariae, but these latter structures will be dealt with later.

The spines of ophiuroids (Fig. 13*i* to *l*) are almost entirely restricted to the lateral plates of the arms. They are very like the spines of asteroids in structure (we are coming to expect this of most ophiuroid features!), with the exception that here, unlike their forbears, the spines are not all straight and fairly smooth but may be hooked or thorny, as in *Ophiothrix* (Fig. 13*l*) and *Gorgonocephalus*, presumably to help in locomotion; or they may be shaped like an umbrella, as in *Ophiotholia* (Fig. 13*j*), possibly for use on softer substrata.

It is among the echinoids that we find the most highly developed spines. In this group, at least in certain members,

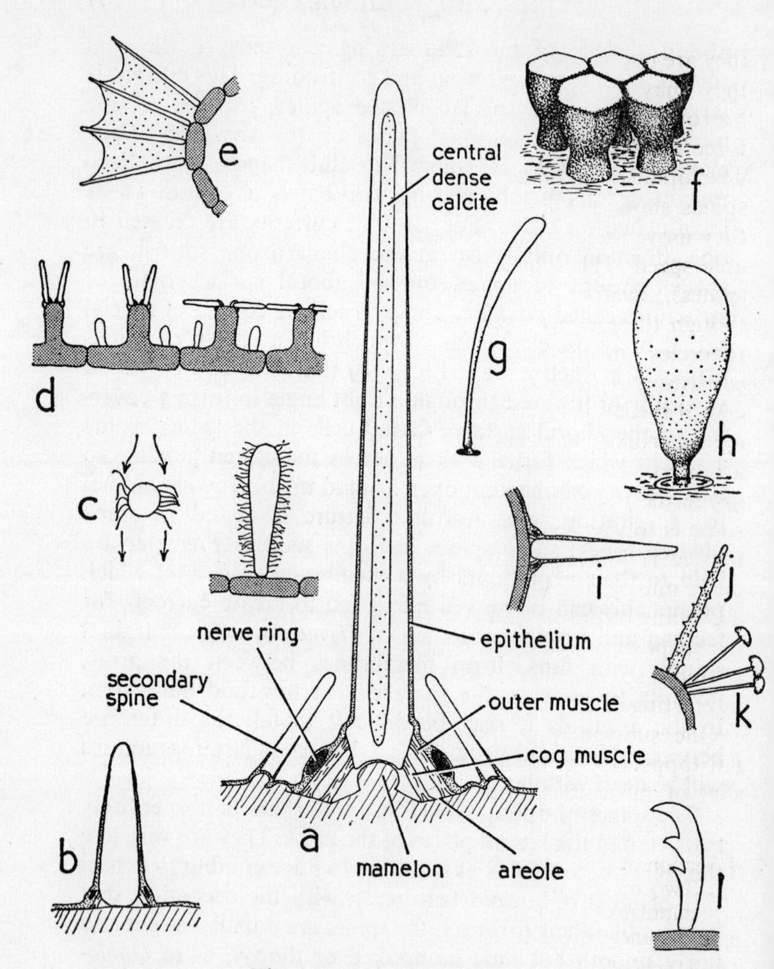

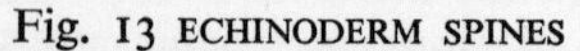

Fig. 13 ECHINODERM SPINES

a Diagrammatic longitudinal section of an echinoid primary spine and the tubercle which bears it.

b to *e*, asteroid spines.

b Simple spine borne on the surface of an ossicle, with no tubercle.

c Side and plan view of a clavula, showing bands of cilia.

they are put to a greater variety of uses than in other groups; they may be used for locomotion, digging, protection, burrow-building, production of currents, breaking the force of waves, bearing poison glands and harbouring the developing larvae. In at least one urchin, *Diadema*, the spines show metachronal rhythm during locomotion, and they move the animal across the ocean floor with considerable speed. The spines of echinoids may be large (*primary* spines), medium (*secondary*) or small (*miliary* spines), though there is no clear division. All are borne on special tubercles on the thecal plates (Fig. 13*a*), each tubercle consisting of a central *boss*, on which the concave proximal end of the spine articulates like a ball-and-socket joint, and a saucer-like *mamelon*, to which the spine muscles are attached. These muscles are in two conical layers, one inside the other, and the inner layer acts as a clamp if the spine is touched; if another part of the test is touched, this muscle is inhibited, and the spine can be moved by the outer muscle. Round the base of each spine is a nerve ring, a swelling of the epithelial plexus. It would be a task beyond the scope of this book to describe all forms and functions of echinoid spines, but the following is a selection of the more interesting ones.

One subtropical echinoid, *Colobocentrotus*, has special flat-topped spines on its aboral surface (Fig. 13*f*) forming a sort of false test to the body, apparently for protection

---

d Section of four paxillae from aboral surface of a phanerozone. Two on the left are open, two on the right closed. Four papulae are included, out of the plane of the section.
e Fan-spines of *Pteraster*.

f to *h*, echinoid spines. See also *a*.
f Flat-topped protective spine of *Colobocentrotus*.
g Paddle-shaped spine of a spatangoid.
h Club-shaped spine, possibly poisonous, of a fossil cidarid.

i to *l*, ophiuroid spines.
i Simple lateral spines.
j Thorny spines, and *k*, umbrella spines of *Ophiotholia*.
l Hooked spine of *Ophiothrix*.

against wave-action; it seems likely that the Triassic *Anaulocidaris* had a similar adaptation[41]. The genus *Cidaris* and urchins close to it have relatively huge spines, mainly used for locomotion. But in a few of them the ends are expanded in a number of rays, the most usual being a club-shaped form bearing poison glands (Fig. 13*h*); a cidaroid having spines of this sort was common in the British Upper Cretaceous, and some levels of the Chalk bear many of these spines, looking like Indian clubs. In the spatangoids some spines are oar-shaped, for pressing against the particles of a sandy or muddy substratum, and also for wiping mucus on to the burrow walls to stop them from falling in. Clavulae also occur in spatangoids. These, closely similar to the spines here given the same name in asteroids, occur in special bands called *fascioles*, diagnostic of the order Spatangoida. These aggregations of ciliated spines produce currents in local parts of the animal's burrow for such purposes as feeding, respiration and excretion.

The last spine adaptation to be described is a very interesting one: all echinoids except cidaroids possess tiny spherical modified spines called *sphaeridia*, each borne on a tiny tubercle usually in a pit in the test. They are generally found in the ambulacral areas around the mouth, but some urchins have them along the entire ambulacra. They appear to be organs of balance, since an urchin which is deprived of them and then inverted will take between five and ten times as long to right itself as an untouched urchin. Their structure has not been adequately investigated, but it appears that the weight of a sphaeridium hanging out of its usual position will cause sensitive areas at its base to be stimulated.

The pedicellariae

When an animal with a hard body is the only solid thing on an otherwise soft sea-bottom the risk of pelagic larvae of sessile animals settling on it must be great. It is probably for this reason, among others, that the two living classes with relatively non-flexible exterior surfaces, the echinoids and asteroids, are provided with almost unique pincer-like organs, the pedicellariae, over their entire exposed surfaces so that the animals can deal with settling larvae before they have a chance to metamorphose into something which could cover such vital structures as the respiratory organs or the tube-feet. Even such relatively hidden animals as the spatangoid urchins in their snug burrows possess these organs, so probably they are not spared the menace of settling larvae. It is interesting that some members of another invertebrate phylum, the Ectoprocta, possess modified polypides, the avicularia, with a structure remarkably convergent on that of the pedicellariae; they are also said to have a similar function.

Of those classes of the echinoderms which are not provided with pedicellariae, the holothurians alone present a large surface area to a potential settler, but ripples of muscular activity are constantly passing down the body, and these would presumably be sufficient to dissuade the larvae. In the case of the other two classes, ophiuroids and crinoids, the only large area exposed to danger of this sort would be the aboral surface of the disk of the one and the tegmen of the other; and both these surfaces are flexible.

Another interesting point is that the tests of many fossil echinoids, particularly those which inhabited the silt at the bottom of calm oceans, are dotted with sessile animals such as ectoprocts, brachiopods, serpulid worms and so on, and these certainly appear to have settled after the urchin's death, because they spread across the tubercles and pores, and even across mouth, anus or apical region, which is hardly likely to happen in a living animal. Here, then, in a calm sea where the bottom is a soft calcareous ooze, an urchin would come to the surface and die, its spines would drop off and it would provide the rare commodity of a hard surface, an oasis for the settler in an otherwise desert expanse of silt.

The general anatomical plan of a pedicellaria is as

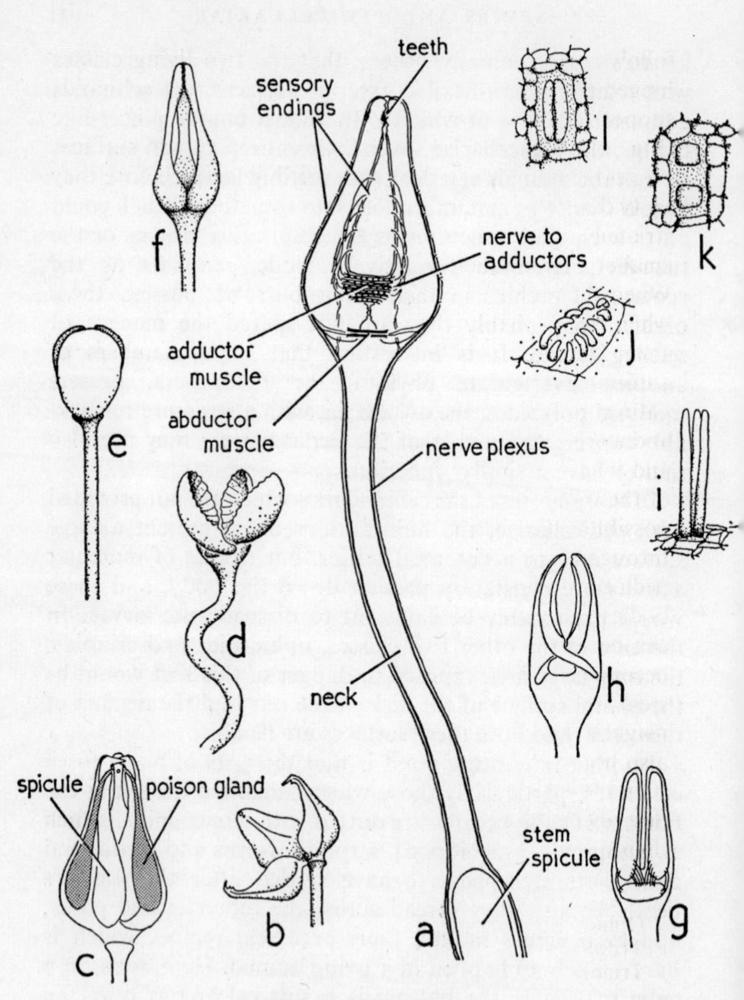

Fig. 14 PEDICELLARIAE

a Section of a typical pedicellaria, based on an echinoid tridactyle.

b to *f*, echinoid pedicellariae.

b Glandular type, open, and *c* closed, in section, showing poison glands.

follows: a number of articulating blades, usually three, hinged proximally, are borne on a movable stem which is supported along at least part of its length by a calcite rod (Fig. 14*a*). At the base of the stem the sub-epithelial nerve plexus is expanded into a nerve ring. The blades are operated by two series of antagonistic muscles, the large, sometimes striated, adductors distal to the hinge for closing and the smaller non-striated abductors proximal to it for opening. When they are touched on the outside, the jaws open, and when they are touched on the inside, they generally snap shut; from this, one infers that they are operated by two autonomous systems involving reflex arcs, the one originating at sense cells on the outside and innervating the abductors, and the other having sense cells on the inside and innervating the adductors. It is curious that the nerves of the outer face have never been described. However, it is possible to see fine processes, probably sensory, protruding through the epithelium on the inside of the jaws of some pedicellariae, and one can follow the nerves past their cell bodies into the adductor muscle; some additional fibres can be seen leading towards the stem, where they probably become incorporated into the sub-epithelial plexus of the stem; this may explain how pedicellariae and spines in the neighbourhood of a pedicellaria which has been stimulated also become active.

In the echinoids four main types of pedicellaria have been described, though they vary enormously within the main divisions. The largest and commonest of these are the

---

d Ophiocephalous, open.
e Trifoliate, closed.
f Tridactyle, closed.

g to *l*, asteroid pedicellariae.
g Straight type, showing muscles.
h Crossed type.
i Sessile type.
j Pectinate type.
k Alveolar type.
l Bivalve type.

tridactyles ('three fingered') (Fig. 14*f*). A calcite rod supports about the first two-thirds of the stem, the head being borne on a flexible neck capable of considerable movement. The blades touch at their tips, where there are two or three tooth-like projections. Modifications of this type, with anything from two to five blades, are found in various groups. The second type is the *trifoliate* ('three-leaved') (Fig. 14*e*), with broad leaf-like blades which meet only along their lateral edges. The third, the *ophiocephalous* ('snake-headed') (Fig. 14*d*), are found chiefly on the peristome, together with trifoliates. The chief features of these are, first, that serrations are present down the whole length of the distal side of each jaw, and, secondly, that each jaw has an inwardly directed process proximally, possibly to give better holding power. The last type, *gemmiform* or *glandular* (Fig. 14*b*, *c*), are provided with two or more teeth distally, to pierce the prey, and poison glands, either in or against the blades or in the neck of the stem or both, to paralyse it. Round the poison glands there may be a muscle sheath, to ensure discharge of the poison at the right time. The behaviour of this type is interesting: like the others, the blades will open when touched on the outside, but will open wider if touched on the sensory processes inside; only if a chemical stimulus is applied to the nearby test will they close and discharge poison, and if the stimulus is applied only to the sensory processes inside, then discharge of poison occurs but the jaws remain open. It is said that if forty of these pedicellariae are boiled in 1 c.c. of water and injected into a rabbit, death follows in two or three minutes; certainly the Japanese do not like to handle one of their native species, *Toxopneustes pileolus*, for long, because of the discharge of poison from their glandular pedicellariae[58].

In the asteroids there are four types of pedicellaria, but they are so different from all those of the echinoids that they have probably evolved separately. Of the four types the first two are borne on a stalk and may be either *straight* (Fig. 14*g*) or *crossed* (Fig. 14*h*), with an accessory ossicle on which the blades hinge and to which the muscles are attached. The third type, *sessile* (Fig. 14*i*), consists essentially of three little-modified spines arising close together on the animal. In the fourth type, *alveolar* (Fig. 14*k*), the moving parts are sunk in the body wall and work by sliding in a plane parallel to the body. It is easy to visualize the starting-point in pedicellaria evolution from ordinary spines when one sees the asteroid sessile type, which must surely represent the simplest arrangement of the series.

Each echinoid and asteroid has evolved to deal with its own special problems of larval and detrital settlement, the kinds of settlers and the nature of detritus being different from those harassing its relatives; so it is not surprising that systematists very often look to the pedicellaria blade structure for a convenient and relatively non-variable specific character. It would be interesting to know the exact function of each type; that is, what problems are presented by the larva or larvae that each can deal with; we know, for instance, that a tridactyle will snap shut quicker than any other type, because of its striated muscle, and that an ophiocephalous will hang on more tenaciously when closed; we know that some trifoliates have been seen to break up bits of sediment so that they can more easily be removed by the cilia. But why, for instance, these latter two types are absent from cidaroids we do not know; neither have we anything like a complete picture of the interaction of the various types of pedicellaria possessed by each echinoid or asteroid in combating one of the most serious menaces of the ocean-floor.

9

THE TUBE-FEET AND THEIR EVOLUTION

Of all the appendages borne by the echinoderms the tube-feet are the only ones which are found in every living class, and probably were present on many of the extinct forms as well. In their delicacy and in their multiplicity of function they are fascinating, and in their combined strengths in some situations they are astonishing. Though strikingly parallel organs are found in many other groups of the invertebrates, for instance in the food-collecting tentacles of some sipunculoids, with their muscular 'compensation sacs', yet as hydrostatic organs the tube-feet must surely take pride of place in the animal kingdom.

These organs exhibit an amazing variety of structure and function; they may be locomotory, tactile or chemosensory, or feeding, burrow-building or respiratory, the latter probably being the original function (p.116). Yet the basic histological plan of all of them is remarkably similar. Externally each has a covering epithelium which is continuous with that covering the rest of the test (Fig. 15*a*). Normally, an integral part of the epithelium is the nerve plexus which underlies the covering cells; where provision is made in a tube-foot for crinkling the surface during contraction, the epithelium and the underlying nerve plexus fold as one layer. The plexus is thickened on one side of the stem to form the longitudinal tube-foot nerve, and also usually at the distal and proximal ends of the stem to form nerve rings. Lining the lumen of the tube-foot is the coelomic epithelium, continuous with that lining the rest of the water vascular system. It is normally ciliated, and in the stems of most tube-feet the cilia are arranged in two longitudinal bands, beating in opposite directions, so that a circulation of coelomic fluid is maintained in the lumen. Between these two epithelia are two important layers: first, next to the external epithelium, is a layer of connective tissue which forms the main structural framework of the tube-foot and may for this purpose contain embedded spicules; and, secondly, the retractor muscles internal to the connective tissue sheath. In addition to withdrawing the tube-foot, these apparently act also in postural bending: it seems that any sector of the muscle cylinder can be contracted independently of the rest (though so far separate innervation to the various parts of the cylinder has not been shown anatomically), and can pull in opposition to the pressure in the opposite direction of the coelomic fluid in the tube-foot lumen to produce flexure.

Distally the tube-feet may be expanded to form a disk. The commonest form of this is as an expanded plate forming a sucker. This has evolved, probably independently, in some members of at least three living groups, the asteroids, echinoids and holothuroids. In the other two living groups, ophiuroids and crinoids, the tube-feet are normally papillate with somewhat pointed ends.

If a tube-foot is to be extended, some method of forcing coelomic fluid into the lumen is required. In the crinoids this is brought about by isolating parts of the water vascular canal and contracting the compartments thus formed so that fluid is pumped out to the tube-feet[69]. In some ophiuroids, too, a somewhat similar method is found, but here, in addition to the isolation of successive parts of the water vascular canal, the side-branch from the canal to a tube-foot, though embedded in the ambulacral ossicle, is expanded to form a muscular chamber for the purpose of protraction. In asteroids, echinoids and holothuroids, where rather greater powers of extensibility are seen, blind-ending muscular *ampullae* protrude into the perivisceral

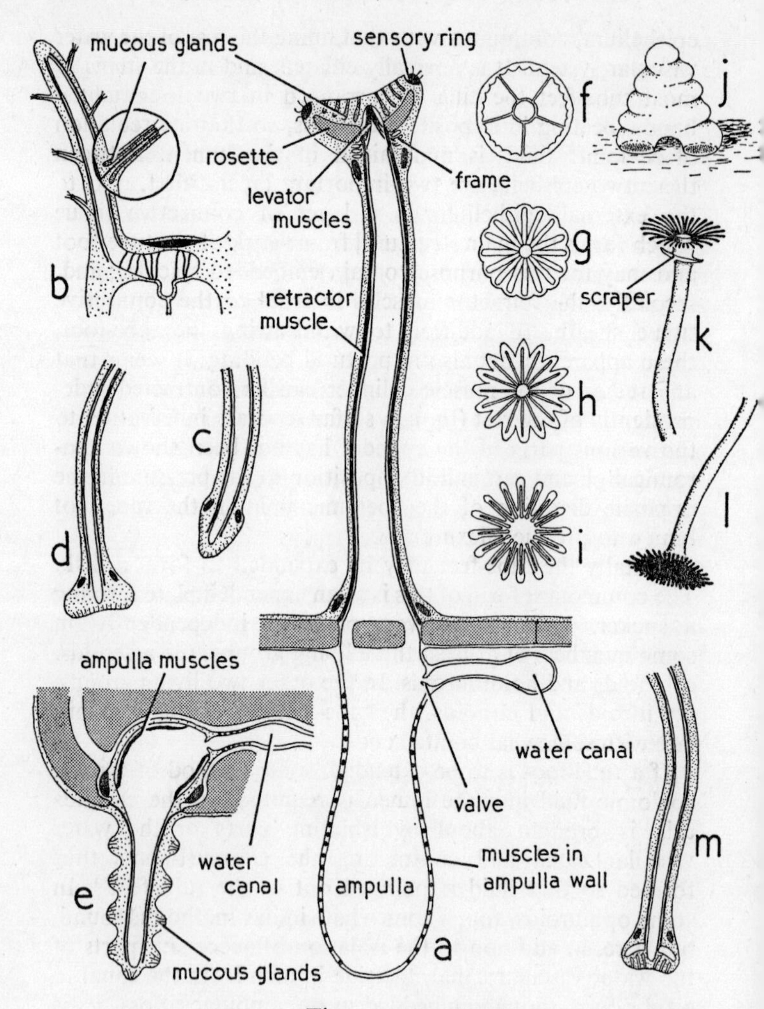

Fig. 15 TUBE-FEET

a Diagrammatic longitudinal section of a suckered tube-foot, based on that of *Echinus*. Skeletal parts and mucous glands densely stippled; epithelial and connective tissues sparsely stippled.

coelom (Fig. 15*a*). Each ampulla consists of little more than a covering epithelium, continuous with that lining the perivisceral coelom, an inner epithelium, continuous with that lining the rest of the water vascular system, and a central sheath of muscle, antagonistic in activity to the stem retractors of the tube-foot; there may also be a thin layer of connective tissue between the muscle and the coelomic epithelium. At the point where the branch from the radial water vascular canal joins the lumen of each tube-foot/ampulla system there is a valve, so that when the muscles of an ampulla contract the fluid in it can pass only into the lumen of the associated tube-foot.

In almost all echinoderms there is some division of labour among the tube-feet of any one animal: in almost all forms a group of tube-feet round the mouth differ from those arising from the rest of the body, and there may be further divisions in those of the rest of the body. The oral tube-feet, which often arise from the circumoral water ring rather than from the radial water vessels, are nearly always

---

b Food-catching tube-foot of a crinoid, based on that of *Antedon*, showing one of the mucus-producing papillae enlarged.

c and *d*, asteroid tube-feet.
c Digging tube-foot of a phanerozone.
d Suckered tube-foot of a cryptozone.

e Tube-foot of an ophiuroid, showing the ampulla-like expanded part of the canal.

f to *l*, echinoid tube-feet. See also *a*.

f to *i*, diagrams of the disks in plan view, showing the evolution of the skeletal structures in penicillate tube-feet of spatangoids.
f Echinus, *g Brissopsis*, *h Schizaster*, *i Echinocardium*.
j Respiratory tube-foot of a spatangoid.
k Funnel-building tube-foot of a spatangoid, showing the excavating scraper and the mucus-producing papillae.
l Feeding tube-foot of a spatangoid.

m Holothuroid suckered tube-foot.

entirely sensory, though in some groups, notably the holothuroids and some echinoids, they may assume an important function in feeding.

Crinoids

Of all the tube-feet of any echinoderm that I have ever seen in activity, those of the comatulid *Antedon* are by far the most vigorous. They are mainly used as feeding organs, with possibly a subsidiary sensory function. The animal sits attached to the substratum by its cirri with its arms held out to form a collecting bowl, its mouth being at the centre. The food-collecting grooves, extending along each arm and pinnule, except the oral pinnules, are lined by tube-feet in groups of three. Along the arms the tube-feet of each group are of equal length, and when extended lie at an angle of about forty-five degrees to the vertical plane of the arm, but along the pinnules the three tube-feet of each group are of different lengths and each lies at a different angle: thus, the largest of each group projects nearly horizontally in the valley between two adjacent lappets (the cover plates which can fold over and protect the groove), the medium-sized ones project at about forty-five degrees and the smallest stick almost vertically upright.

The external epithelium of each tube-foot is drawn out at intervals into finger-like papillae (Fig. 15*b*). Each papilla is solid, with a single muscle fibre at the centre running its entire length; no other muscles are present in the epithelium. At the base of each papilla are five or more single-celled mucous glands with long ducts passing up the length of the papilla to open at its distal end, and a group of nerve cells which send sensory processes to protrude beyond the end of the papilla. Watching *Antedon* feed, one gains the impression that the tube-feet are stimulated into their violent, flailing activity by the touch of food on the sensory ends of their papillae. These then possibly cause the central muscle to contract, which squeezes mucus from the glands, and at the same time the tube-foot bends violently, the effect of this being to throw out a series of mucous strings in which the food is trapped. Now the task is to get these strings safely into the food grooves, where the cilia will transport them to the mouth. This is where the three sizes of pinnular tube-feet come in: the largest are clearly the main food-catching ones, and these get rid of their food-laden strings by throwing them to the medium-sized ones, which bend both outwards to pick up the strings and inwards to pass them to the smallest ones, and these bend in all directions to push the strings into the food groove. This differential activity is reflected by the retractor musculature of the three sizes of tube-foot: in the largest the muscles are on the oral (uppermost) side of the stem only, in the medium they are on the oral and aboral sides only, while on the smallest they are all round[69].

Asteroids

Two main types of tube-foot are found in this class: pointed in those which burrow (mainly phanerozones) and suckered in those which move over the surface of the substratum. In the pointed ones (Fig. 14*c*) the connective tissue sheath is expanded at the end into a strong arrow-head for thrusting into the substratum during burrowing, and the epithelium at the tip is heavily charged with mucous glands. These tube-feet probably have a dual function: they assist in feeding by helping to move particles along the ambulacral grooves towards the mouth and they probably also extend up the lateral sides of the arms during burrowing, even possibly across the oral surface, to plaster mucus on to the burrow walls and prevent them falling in.

The connective tissue at the distal end of the suckered tube-feet (Fig. 15*d*) is also expanded to form the main framework. Smith [76,77] has shown how the forces exerted by the retractor muscles during adhesion tend to maintain the full diameter of the sucking disk by pulling on the most peripheral of the connective tissue strands. Here again the disk is well supplied with mucous glands, to help in adhesion.

There is an interesting anatomical difference between the burrowers and the non-burrowers: in some of the burrowing asteroids each ampulla has two separate lobes to it, whereas the ampulla of the suckered tube-feet is a simple sac. There are two possible explanations of the bilobed arrangement: first, that each lobe is separately innervated and only during burrowing activity, when more extension is required, is the second lobe activated; or, secondly and more probably, that a single ampulla has been secondarily subdivided to provide greater surface-to-volume ratio, and hence more muscles in the ampulla wall, so that during burrowing the tube-feet can make powerful thrusts into the substratum.

Echinoids

The recent regular echinoids have the most highly developed suckers on their tube-feet of any class. The sucker disks are supported and assisted in their activity by a fairly complex series of calcareous ossicles, which functionally replace the rather dense connective tissue plate of the asteroid suckered foot[70]. At the distal end of the stem there is a complete ring of ossicles, the *frame*, which lies on the proximal side of a much larger skeletal structure, the *rosette* (Fig. 15*a*, *f*). The function of the frame is to act as an anchor for a special set of muscles. The rosette has two functions: first, it helps to keep the shape and width of the sucker during adhesion, when all the forces acting to raise the diaphragm are tending to pull in the disk edges, and, secondly, it operates during detachment by transmitting the pull of the stem retractor muscles to the edge of the disk, which is the most effective place to apply a pull for releasing the sucker. The stem retractors are inserted at the connective tissue sheath adjacent to the inner border of the rosette to be effective in this way. Within the lumen is another set of muscles, the *levators*, which originate at the level of the skeletal frame and run between the retractor muscle fibres before passing distally to attach at the centre of the disk. Between the rosette and the disk surface are large multicellular mucous glands to help in adhesion, and round the periphery of the disk, at least in the more advanced regular echinoids, is a sensory ring.

As far as one can tell from the fossil record, the dramatic change from regular to irregular, and the consequent potential for exploiting new habitats, occurred more than once: the holectypoids and clypeasteroids arose independently of the spatangoids. The first two orders are included in Zittel's super order Gnathostomata (with a jaw apparatus), which suggests rather less divergence from regular echinoid structure. This is true of tube-foot structure too; though there appears to be no detailed histological study of the tube-feet of any of the three holectypoid species still extant (two species of *Echinoneus* and one of *Micropetalon*), Lovén[42] gives a picture of the skeletal structures in the disk of *E. semilunaris* showing these to be similar to those of regular urchins. In clypeasteroids, too, the tube-foot structure of at least one, *Echinocyamus*, appears to deviate very little from the regular echinoid plan[68], with the exception that the skeletal elements of the disk are missing and are functionally replaced by epithelial muscles which apparently serve to raise the outer parts of the disk relative to the centre during detachment.

So far only the suckered tube-feet occurring over the lower parts of the test have been described. In addition to the difference between these and the wholly sensory group round the mouth (p. 109), there is further division of labour in each ambulacrum in that aborally (i.e. nearer the apical disk) the tube-feet tend to lose their suckers and become, apparently, mainly respiratory and sensory in function.

In the spatangoids, where the tube-foot pattern is drastically changed because of the highly developed burrowing habit of most of them, the demarcation of the solely respiratory tube-feet is much more abrupt. In addition, the suckered tube-feet are no longer required, and they are functionally replaced, in appropriate positions on the animal, by burrow-building, sensory or feeding tube-feet.

Thus, taking the common British heart-urchin *Echinocardium* as an example[67], the tube-feet in the dorsal parts of four of the five ambulacra are respiratory, those in the dorsal part of the fifth have the function of building a special respiratory funnel from the burrow to the surface of the substratum, the feet round the mouth are for feeding, some on the posterior side build a special sanitary drain, and those round the ambitus (the widest part of the test) and in parts of the oral surface are for sensation only.

The surface area of the disk of mucus-producing tube-feet is normally vastly increased by the presence of numerous papillae (Fig. 15*k*, *l*). The disk in these tube-feet is not supported by a rosette, but each papilla has a single spicule supporting it, at least along part of its length. One can trace a morphological series in the papillae and their skeletal elements from the typical regular echinoid condition (Fig. 14*f*), through the funnel-building tube-feet of such brissid spatangoids as *Brissopsis* (Fig. 15*g*), in which the disk edge is merely scalloped and the fifteen or so spicules form a structure very like a rosette, touching centripetally, through *Schizaster* (Fig. 15*h*), where the papillae are more marked and the spicules reduced, to *Echinocardium* (Fig. 15*i*), where there are two rows of papillae and the spicules are reduced to mere rods and do not touch at their inner ends. In *Brissopsis* and *Echinocardium*, and probably in the others as well, the stem retractors are inserted at the inner border of the skeletal elements, which suggests that they are homologous with the elements of the regular echinoid rosette. As there is no levator muscle system in the spatangoid tube-feet (there being no sucker), there is no need for a skeletal frame as such. But in the funnel-building tube-feet of *Echinocardium*, for instance, and in some others as well, one (or occasionally two) very large spicules may be present on the side of the disk (Fig. 14*k*) in a position roughly corresponding to that of the frame of regulars. It seems most likely that the purpose of this is to act as a scraper for the walls of the funnel, the sand removed by them during excavation being taken down to the cavity occupied by the animal and passed to the rear of the burrow, where it fills in the space formerly occupied by the animal as it moves forward. A great deal of mucus must be required to plaster the walls of the funnel excavated in this way, so it is not surprising to find that the epithelium covering the disk and papillae of the tube-feet contains numerous mucous glands. Further, the epithelium has muscle fibres running between the glands, most probably for more efficient discharge, and to ensure that the mucus is cast on to the funnel walls and does not remain on the tube-feet. A comparable arrangement is found in those tube-feet of spatangoids which build and maintain the sub-anal sanitary drain, but it is interesting that in the oral feeding tube-feet, superficially similar to the burrow-building ones, the glands are *not* provided with muscles, because the mucus is required to remain on the tube-foot.

Ophiuroids

The tube-feet of the brittle-stars have received far less attention than those of other classes. In general, they show a structure rather similar to that of the crinoids in that most of them are slender and pointed, and the epithelium down the entire length of the stem is raised at intervals into papillae (Fig. 15*e*), containing mucous glands and sense organs. As indicated above, some ophiuroids, notably burrowers, have incipient ampullae and valves in the canals leading to their tube-feet.

Holothuroids

These echinoderms, lying on their sides, show, typically, a division of labour chiefly because of the adoption of a locomotory function by the tube-feet of the three ventral ambulacra and the reduction of those of the other two to small sensory papillae. This, of course, does not hold good fort he Apoda (p.87) and the other burrowing forms. In addition, the oral tube-feet, even in the Apoda, are usually very large relative to the body size, and may even have an extended length equal to half the length of the body. The wide variety of structure in these organs is used extensively by taxonomists, and indicates that a correspondingly wide variety of feeding methods is employed by the class. Typically, the oral feeding tube-feet are dendritic, the stems being highly muscular and the ends of each branch being richly supplied with mucous glands and sense cells. When the animal is feeding, these organs are extended in a sweeping motion, apparently scavenging particles of detritus from its neighbourhood, then they are bent round into the mouth while retracting somewhat, and finally they wipe their mucus with its collected food off against the oral ring of ossicles and extend again.

The suckered tube-feet (Fig. 15*m*) show a pattern which is different again from that of the asteroid and echinoid suckers. Here there is a single, slightly dome-shaped spicule as a support, with a ring of very minute arcuate spicules on its proximal side, peripherally. Groups of unicellular mucous glands lie distal to the skeletal plate, and connective tissue fibres pass through its fenestrae to be inserted at the disk cuticle. How the sucker functions is not known precisely, but it is probable that the stickiness of the mucus is largely responsible for adhesion.

The evolution of the tube-foot/ampulla system

So rarely are the remains of tube-feet and associated structures left behind in the fossil record that we know very little of the origin and early history of this system. The first echinoderms to appear, simple crinoids (p. 29), tell us nothing in this respect. One can suggest that at first exchange of gases occurred across the walls of the simple ambulacral canal, mentioned in Chapter 1, of the ancestral echinoderm, and that later the efficiency was increased by having side-branches, the tube-feet. It is not unreasonable to imagine their walls becoming muscular to assist in the collection and despatch of food, which is the situation one finds in modern crinoids. But the stages by which this was attained will very likely never be known.

There is, however, one piece of indirect evidence which suggests that the crinoid tube-feet arose at least before the Upper Cambrian, since by this time the first agelacrinid edrioasteroids (p. 149) had appeared, and, from the evidence of pores in the cover-plates to their ambulacra, we can be fairly sure that they had side-branches to their ambulacral canals which waved free in the surrounding water.

Little is known about the relations of the early echinoderm classes, so one cannot say how the respiratory organs of the non-crinoid pelmatozoa fit in with this. The facts suggest that the cystoids (p. 133), blastoids (p. 141) and heterosteles (p. 155) lacked any connected system of protrusible organs, although ambulacral canals were certainly present in the blastoids and may well have been present in the others also, underlying the epithelium of the ciliated food grooves to bring nutrients and oxygen to the cilia. In all these forms respiration appears to have been carried out by soft structures which arose from scattered isolated pores piercing the theca, such as the diplopores and rhombopores of the cystoids, hydrospires of the blastoids and the 'gill-slits' of the heterosteles. Certainly, in some later cystoids, such as *Proteroblastus* (Fig. 19*f*), the diplopores tend to be restricted to the plates adjacent to the food grooves, as though deriving benefit from the steady current in that region, and the same may be said of the blastoid hydrospires. But such systems of respiratory organs were entirely separate from the vascular canals under the food grooves, where these occurred, and may well have suffered from the disadvantage that the oxygen obtained by them diffused into the perivisceral coelom rather than into a vascular system. This may explain the greater degree of success of those echinoderm groups with tube-feet.

But, to return to the main line of the echinoderms which led to the eleutherozoa, we have already said that the early edrioasteroids probably had tube-feet by the time they appeared in the record, but so far no echinoderm has shown any sign of ampullae. Evidence of these first appears in the Ordovician, with the arrival of the non-agelacrinid edrioasteroids such as *Edrioaster* and *Pyrgocystis*. These forms had pores between the ambulacral plates to carry the canal from the tube-feet to their ampullae internal to the skeleton. The presence of ampullae immediately suggests that the tube-feet were extensile and hence muscular, but whether they could be bent to assist in feeding we do not know. As soon as the Eleutherozoa proper arose and turned 'face downwards' or 'face forwards' the possession of organs which could be used in locomotion would be most advantageous. Possibly at first they relied on the stickiness of their mucus for attachment, as do most modern ophiuroids, but the tube-feet would be much more efficient under some circumstances if provided with a sucker; and this is the pattern exploited by the other eleutherozoans, the asteroids, echinoids and holothuroids.

10

LARVAL FORMS AND THEIR METAMORPHOSIS

WHILST radial symmetry is one of the characteristic features in the adult echinoderm and an attached condition is very widespread, in the larval state the various classes share a bilateral plan, and are invariably free for most of their life. We can safely assume that the extinct groups also developed indirectly, through larval forms which were probably very similar to those of living echinoderms. A larva has a double task to perform: to distribute the species and to grow up into the adult. In the echinoderms, as much as in any other animal group, we can see that larval evolution has taken place independently of that of the adult, and is subject to conspicuous specialization, even among closely related types; for example, among the echinoids *Eucidaris* has a planktotrophic larva (feeding on plankton), a free-floating form with long arms, whereas *Heliocidaris* has a lecithotrophic larva (feeding on stored yolk), which has the shape of a barrel[87]. Some recent groups even dispense with a larval stage altogether: that is, they show direct development, usually but not always coupled with the possession of rather more yolky eggs and with some form of marsupial or incubatory care of young by the parent (p.130); but this is the exception, and in most cases the eggs and sperm are shed into the sea, and fertilization and development take place there.

Early development

The egg of most echinoderms is a sphere of about 75 μ in diameter surrounded by a jelly coat; the sperm is about 50 μ in length, with a pointed conical head, a middle-piece the shape of a flattened cylinder containing two centrioles, and a long thin tail. There is an acrosome, like an arrowhead, at the tip, and it has been said that there is also a pore through which a sticky substance is secreted as the sperm meets the egg.

Cleavage of the echinoderm egg is total, indeterminate and radial. The resulting blastula is oval, hollow and ciliated all over. An inpushing at one end forms the beginnings of the larval gut, or *archenteron*. The hole through which the archenteron opens to the exterior is the *blastopore*. The blastopore marks the position of the future anus: that is, the echinoderms are *deuterostomes*. There now follows a stage of development which is considered highly important from the point of view of phyletic relationships: coelom formation. It is also a stage in which considerable variation occurs, which makes it all the more difficult to interpret. In the vast majority of the echinoderms it happens by *enterocoely*; that is, the coelomic sacs begin as outpushings from the wall of the archenteron, one on each side. But in a very few forms, notably a few species of ophiuroids, it develops by the other method, *schizocoely*; that is, by splitting of the mesoderm, a method more usual in those invertebrates with spiral cleavage, such as the annelids. This latter type of cleavage is also the method seen in the vertebrates, and its occurrence in a few echinoderms has been considered by some authorities[127, 128] to be important with regard to the ancestry of the chordates.

Almost as soon as the primary coelomic sacs have been formed, they bud off posteriorly another pair of sacs, the *somatocoels*, later to form the main coelom of the adult body; then the right anterior sac usually degenerates, while the left one becomes bilobed. This is called the left *axohydrocoel*, the anterior part being the *axocoel* (most of which degenerates in the adult) and the posterior the *hydrocoel* (later forming the cavities of the water vascular system). The left axohydrocoel now forms a connexion with the exterior by a small canal leading to a hole on the dorsal surface, the *hydropore*. Meanwhile, an inpushing on the ventral surface pierces the body wall to give rise to the larval mouth.

The dipleurula

It is not strange that early in life a common plan exists between those forms with larvae, or most of them. This stage has been termed the *dipleurula* ('little two sides'), a word which has unhappily been used to denote also a hypothetical common ancestor to the entire phylum (p.175). The main thing about the dipleurula is that the cilia, which in the early embryo covered the entire body, now aggregate into a single band forming a closed loop which starts just in front of the mouth, passes up the sides and along the top, then down and under the larva just in front of the anus (Fig. 16*a*). This, then, represents the basic larval type from which most of the various later larvae are derived. No known echinoderm has only a dipleurula before metamorphosis: all are subsequently modified for their own particular larval conditions. We shall now consider the larvae of the various classes in turn, and how they are derived from the basic pattern.

The holothuroids

The simplest situation is probably that seen in the holothuroids, in which the dipleurula is transformed into an *auricularia* by the ciliated band becoming sinuous (Fig. 16*g*). In this case the complement of coelomic sacs is the same as in the dipleurula, namely, left axohydrocoel, left and right somatocoels. Before metamorphosis into the adult, the sinuous ciliated band reorganizes by breakage, etc., into three or four rings round the barrel-shaped body (Fig. 16*h*); the larva is now called a *doliolaria*. In some forms, such as

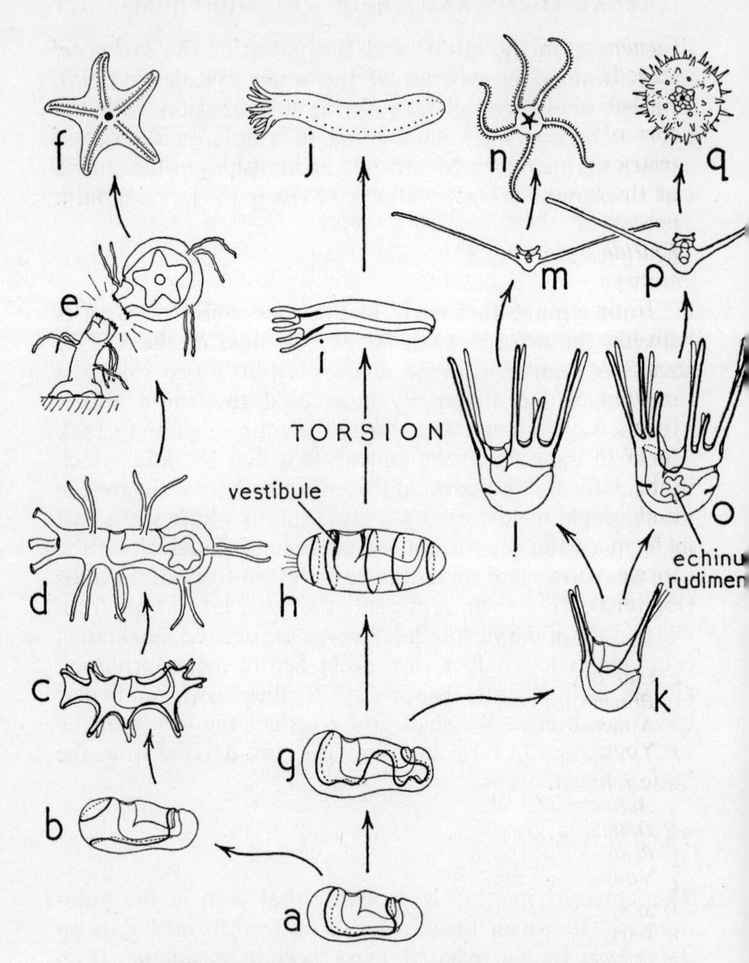

Fig. 16 LARVAL TYPES OF THE ELEUTHEROZOA

The asteroid, holothuroid, ophiuroid and echinoid larval types can all be traced from a basic dipleurula. Thick lines are ciliated bands.

a Dipleurula, side view.

b to *f*, asteroid larvae.

b Early *bipinnaria*, ciliated band divided into two.

Cucumaria planci and *Leptosynapta inhaerens*, the auricularia stage is omitted, while in a few other species of *Cucumaria*, e.g. *saxicola* and *frondosa*, the egg develops into another sort of larva altogether, which has cilia all over it, not restricted to bands. As in other echinoderms, some members of the holothuroids brood their young, but this does not necessarily mean that they dispense with a larval stage: in *Chiridota* a doliolaria stage is passed through by the embryo while incarcerated in the parent's coelom.

In the doliolaria the beginning of the gut, just inside the slit-like mouth, has a thickened wall, and this is called the *vestibule*, a structure which is important in the later development of the oral region. The middle part of the axohydrocoel forms a five-lobed ring round the vestibule, destined to form the five primary oral tube-feet, followed by a sixth lobe, later becoming the polian vesicle; five canals are then budded off, which will become radial water vascular canals. The opening of the hydropore now disappears in those forms with no madreporite in the adult. Five cavities, destined to become the epineural sinuses, originate by

---

c Late *bipinnaria.*
d *Brachiolaria*, with adhesive disks at anterior end.
e Attached *brachiolaria*, star rudiment breaking off.
f Young star.

g to *j*, holothuroid larvae.
g *Auricularia.*
h *Doliolaria.*
i *Pentactula.*
j Young cucumber.

k to *n*, ophiuroid larvae.
k Early *pluteus* with four arms.
l *Ophiopluteus* with eight arms, one pair of epaulettes. Spicules omitted.
m Late *ophiopluteus* from front.
n Young brittle-star.

k to *q*, echinoid larvae.
o *Echinopluteus* with twelve arms, two pairs of epaulettes.
p Late *echinopluteus*, showing similarity to late ophiopluteus (*m*).
q Young urchin.

inpushing of the vestibule wall, and at the same time a ring-like mass of nervous tissue, the future oral nerve ring, arises from the vestibule wall too. The vestibule and the structures which are forming from it now move towards the anterior end of the larva, and the vestibule wall becomes the buccal membrane. At this stage the blastopore closes off, but the anus opens again a little way from it, a curious but fairly common phenomenon of development in deuterostomes which is not fully understood. The lobes of the hydrocoel now push through the buccal membrane and form the lumina of the five primary buccal tube-feet. Usually at about this time, too, two or three tube-feet, the terminal tentacles of the trivium, are formed near the posterior end of the body. The larva at this stage is called a *pentactula* (Fig. 16*i*) and now it settles on the bottom. The adult form is gradually acquired by the growth of other tube-feet over the body and the development of internal organs, such as the respiratory trees from the cloaca. This is a necessarily brief and simplified account of holothuroid development; in general, however, it can be seen that the main parts of the larva are retained in the adult; this is not true of some of the other groups, such as the next.

The asteroids

Very early in larval life in this group the single ciliated band of the dipleurula divides into two, a small pre-oral band and a larger post-oral one. This larva is called the *bipinnaria* (Fig. 16*b*). The development of the asteroid coelom varies somewhat from the general pattern described above. The original pair of pouches, in addition to budding off a pair of somatocoels posteriorly, grow forward and fuse in front to form a U-shaped cavity enclosing the front part of the gut. Most of the accessory coelomic structures, that is, cavities other than the ordinary body (perivisceral) cavity, derive from the left side of this sac, as is normal, but some structures may form from the right side, as will be mentioned later. Both ciliated bands become sinuous, and in some asteroids, such as some species of *Asterina* and *Astropecten*, metamorphosis occurs without further change; but in others, such as *Asterias*, the sides of the larva become drawn out into a number of processes, the contours of which are followed by the ciliated bands (Fig. 16*c*). Next, in most asteroids, the anterior end of the larva becomes drawn out into three arms, each provided with a terminal adhesive disk and with a fourth between their bases; extensions of the axocoel pass into these arms. This larva is termed a *brachiolaria* (Fig. 16*d*). The subsequent development of coelomic and accessory structures has been well worked out[27]. Suffice it to say here that the central part of the left axohydrocoel buds off a five-rayed ring, later to become the circum-oral water ring and the lumina of the five primary tube-feet, and a stone canal develops from this ring to the hydropore. At this stage the remains of the right axohydrocoel in some forms are said to give rise to the dorsal sac, though in others this structure is thought to arise from mesenchyme at the anterior end of the body, while in still others it is said to arise by invagination of ectoderm; the sac has been seen to undergo pulsations, apparently to maintain circulation in the larva.

After about two months at the brachiolaria stage the larva attaches by its adhesive organs, and after a further short time the young starfish which has been forming on the left side of the larva breaks free by a rupture which cuts off the larval mouth from the rest of the gut (Fig. 16*e*). A new mouth pushes in through the water ring and joins up with the larval stomach. Similarly, a new anus is formed opposite the new mouth, that is, on the right side of the larva, which is now the aboral side of the young starfish. There has been a change of symmetry at metamorphosis from bilateral to radial.

The echinoids

Next in complexity come the *pluteus* larvae; this term, originally coined by Müller in the 1850's, means 'easel', but unhappily this refers to the appearance of the larva when upside down! Both echinoids and ophiuroids have this type of larva, distinguished by the appropriate prefix, i.e. *echino*pluteus and *ophio*pluteus. The main feature of the plutei is the great extension of the larval arms, the legs of the easel; these are supported by calcareous spicules, later discarded. In the early echinopluteus two pairs of arms form first, one pair in front of the mouth and the other in front of the anus (Fig. 16*k*), then a further four pairs appear between them (Fig. 16*o*). The main ciliary band (single) follows the contours of the arms, but two ciliary patches, the *epaulettes*, appear at each end of the larval body; these are said to be the main locomotory organs of the larva. Sometimes, particularly in the irregulars, a process, the *spike*, projects downwards from the aboral side, possibly to keep the larva upright.

The process of metamorphosis has been carefully worked out for numerous species. In brief, the common features of it are as follows: the left axohydrocoel forms a five-lobed sac on the left side of the gut, and adjacent to this an in-pushing of the left side of the larval body occurs, forming a vestibule, the entrance of which later closes over. Inside the vestibule the primary podia grow out from the lobed hydrocoel, the epineural canals fold in from the wall and the nerve ring also forms from the wall, as in holothuroids. Meanwhile, the left somatocoel has surrounded the hydrocoel and produced buds between the lobes which will eventually give rise to the elements of Aristotle's lantern. Calcareous elements begin to be deposited. The developing material is now concentrated into a spherical disk on the left side of the larval gut, a disk which the original describers called the *echinus rudiment*; this name has been retained, even for other genera. At metamorphosis the vestibule wall breaks open so that the primary podia can emerge, the coverings of the larval arms are resorbed and their spicules discarded, and a new mouth and anus are formed on the left and right side of the larval body respectively. After the break-through the young urchin, 1 mm or so in diameter, sinks to the bottom.

The ophiuroids

The *ophiopluteus* develops from the dipleurula stage in very much the same way as the echinopluteus, but has only four (sometimes three) pairs of arms finally (Fig. 16*l*), all with skeletal supporting rods. One pair of swimming epaulettes may also appear. Internal development, however, does not follow the echinoid plan. The hydrocoel, instead of developing on the left side of the gut, forms a ring round the oesophagus, and the larval mouth forms the adult mouth; the larval anus closes over, there being no anus in the adult. No vestibule is formed, though the primary podia, when they develop from the hydrocoel ring, emerge into the gut near the mouth, which is equivalent in position to a vestibule. The nerve ring develops simply as an ectodermal swelling and becomes covered by an epineural sinus through the overgrowth of folds of epidermal tissue. The larva continues to swim around during metamorphosis; the larval arms are resorbed and the rods discarded, as in echinoids, and calcareous elements laid down in the body. Then, as the larva settles, the future arms start to grow out from its sides.

The crinoids

This group remains somewhat apart from the others because the gastrula develops direct into a *doliolaria* (Fig. 17*a*), not unlike that of the holothuroids. This has four or five bands of cilia and an adhesive disk somewhere on or near the first band; there is an apical sensory tuft of cilia at the front end. An archenteron forms by invagination but here the blastopore closes off completely, so that the archenteron is now a closed sac lying in the blastocoel, that is, the cavity of the blastula. The archenteron eventually divides into two somatocoels, a hydrocoel, an axocoel and an enteron. There now appears a depression somewhere between the first and third ciliated bands, which will become the vestibule.

All the above structures are formed while the embryo is still in the egg membrane; it now escapes, and remains in the plankton for a matter of hours, or at most days, before selecting a suitable spot with its apical tuft and settling on its adhesive disk (Fig. 16*b*). The vestibule, by now quite a large sac, becomes cut off and the whole inner mass rotates through ninety degrees, bringing the vestibule from the ventral side of the larval body to the future oral (upper) side (Fig. 17*c*); between the vestibule and the enteron the hydrocoel forms a five-lobed ring and then connects with the axocoel, which in its turn has made a connexion with the exterior by a hydropore. The vestibule now sends a projection through the hydrocoel ring to join the enteron (Fig. 17*d*) and this is the future mouth, while the lobes of the hydrocoel grow up into the vestibule as the forerunners of the primary tube-feet. At this stage the larva is called a *cystidean*, because of its supposed similarity to a cystoid. There is no anus as yet, but the gut grows round from the right side of the enteron to the left in a loop (Fig. 17*e*), later to pierce the side of the larva and form the anus, but this will move to the oral surface as development proceeds. The roof of the vestibule is strengthened at this stage by five interradial spicules, the *deltoids*, and soon radial grooves, which later become slits, appear between the deltoids, so that the roof is made up of five short arms. These are not the future adult arms, because the adult arms develop from a second, radial series of plates below the vestibule. The deltoids now bend upwards to permit the primary tube-feet, and later others, to emerge. At this stage the larva is termed a *pentacrinule*, because of its supposed resemblance to the fossil *Pentacrinites*. The plates of the theca are laid down in its walls. The hydropore now closes over, but to replace it many pores develop in the hydrocoel (now water vascular) ring, and others in the floor of the vestibule (now the tegmen). In the comatulids several months may be passed through in the pentacrinule stage, but then the animal breaks its stem at the top and thereafter leads a free existence.

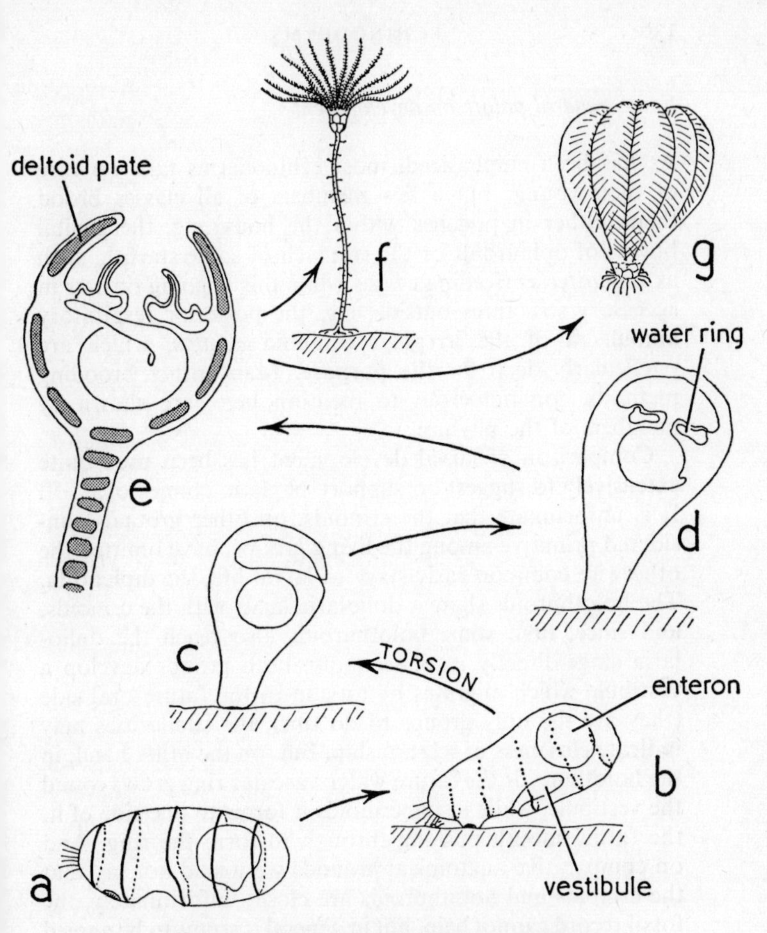

Fig. 17 LARVAL DEVELOPMENT IN CRINOIDS

Crinoids do not pass through a dipleurula stage.

a *Doliolaria*, with enteron already cut off inside.
b Settled *doliolaria*, vestible enlarging.
c *Cystidean* after torsion of vestibule.
d Late *cystidean*, water ring surrounding future mouth.
e *Pentacrinule*, with tube-feet forming, and gut growing round towards left side.
f Adult fixed crinoid.
g Adult comatulid, after breaking from larval stem.

Some general points on development

As has been emphasized, most echinoderms take no care of their young; but a few members of all classes brood them, either in pouches within the body, e.g. the genital bursae of ophiuroids or the stomach of some starfish, such as *Leptasterias* (feeding ceases while this is going on), or in accessory structures outside, e.g. the posterior respiratory ambulacra of the irregular echinoid *Abatus*, which are particularly deep for the purpose. Many other brooding methods, too numerous to mention here, are shown by members of the phylum.

Comparison of larval development has been used quite extensively to suggest or support phyletic connexions[7, 128]. It is unfortunate that the crinoids, on other grounds considered primitive among the living groups, have omitted the otherwise common early stage of larval life, the dipleurula. The holothuroids share a doliolaria larva with the crinoids, and since, first, some holothuroids also reach the doliolaria stage directly, and, secondly, both groups develop a vestibule which migrates by torsion to the future oral side (they are the only groups to do this), the similarities may indicate closeness of relationship. But, on the other hand, in the holothuroids the future water vascular ring grows round the vestibule, while in the crinoids it forms to one side of it, the future mouth pushing through to form the ring. And on comparative anatomical grounds we would not say that the crinoids and holothuroids are close; unfortunately, the fossil record cannot help, but in general it seems to be agreed that their similarities of larval type are convergent.

The rather primitive position of the asteroids, suggested by some features of their adult anatomy (p. 168), is thought to be confirmed by the fact that the brachiolaria larva attaches by its anterior end for the latter part of its life in very much the same way as the cystidean of crinoids; one would expect a difference in subsequent development, however close they are phyletically, because of the different positions of the mouth, and this is shown by the fact that in crinoids there is torsion to bring the mouth uppermost, and in the asteroids there is breakage so that the starfish can settle mouth downwards.

Of more general phyletic interest is the nature of the coelom. The universal presence of three sacs on the left side and the remains of three others on the right suggest a rather close connexion between the echinoderms and other lesser phyla of the coelomates with three pairs of coelomic sacs, such as the chaetognaths, the pogonophores and the hemichordates. Of these, the latter group seems to be the closest relative[118], principally because some have tentacles with coelomic cavities from the central sacs (hydrocoel of the echinoderms, mesocoel of the hemichordates).

One very interesting common feature is possessed by all early echinoderm larvae: the hydropore. Almost nothing is known about its function in the larva; it has been thought of as an exit for accumulated waste from the coelomic pouches—a sort of metanephridium, in fact—or as a pressure equalizer, but it connects only the left axohydrocoel with the exterior, the somatocoels remaining closed. Much more will have to be known about larval ecology and the mechanical problems of their development before the function of this pore becomes clear.

Another common developmental feature of the phylum is the arrangement of those body plates which form first. This follows a fairly constant pattern in all those with an extensive skeleton, that is, all living classes except holothuroids. Usually the first plates to form are a ring of five interradial plates (*genitals* of echinoids, *interradials* of asteroids, *basals* of crinoids), a ring of five radial plates outside and between them, and a central plate (*centro-dorsal* of crinoids, *central* of asteroids, *suranal* of echinoids). Then, in the stellate forms a further radial ring, the *terminals*, is laid down. Hyman[7] does not agree that these plates in the crinoids are homologous with those in other groups, but her criticism appears to be based on the time of development of the cycles of plates only. Actually, it is the ophiuroids which depart most from the basic plan, having the interradials outside the radials, though their homology is established by the fact that one of them forms the madreporite; clearly, their position and subsequent migration to the oral surface is secondary. I have discussed on p.89 the possibility that the unique and constant arrangement of plates may have something to do with the determination of pentamery in the group.

11

EXTINCT FORMS: THE CYSTOIDEA AND BLASTOIDEA

IN THE next two chapters we shall consider those groups of the echinoderms which are entirely extinct. The cystoids and blastoids, to be dealt with first, form a distinct side-line which most likely left no descendants; in the next chapter are included the rest of the extinct forms, some of which are regarded as occupying important positions in the phylogeny of later groups, others as blind early offshoots, and others as enigmatical fossils which may not be echinoderms at all.

The class Cystoidea used to be a veritable 'rag bag' of extinct forms and included all the non-crinoid Pelmatozoa that could definitely be assigned to the phylum, that is, the cystoids as defined here, the group now separated as the Heterostelea, the blastoids and a cystoid-like group, the Aporita, now usually referred to the eocrinoids (p. 29). Nowadays, it is more usual to restrict the cystoids to those non-crinoid pelmatozoa whose plates are pierced by special pores which bore structures apparently independent of any internal system of coelomic canals, such as a water vascular system, and hence are not to be regarded as tube-feet. The theca (Fig. 19*a*) is flask-shaped with the mouth in the centre of the upper surface and the anus to one side of the mouth, surrounded by a circlet of five plates, even in the forms otherwise devoid of pentamery (p. 89). Between the mouth and anus are usually two other pores, probably a gonopore (nearest the anus) and a hydropore. Some cystoids, particularly the later forms (Silurian and Devonian), were borne on a stem, which in life was inserted into the substratum by a group of rootlets; but more often the theca sat directly on the sea-bottom. Two apparently independent trends can be followed within this class, one being the change from a haphazard thecal plate arrangement to a more regular, and usually pentamerous, plan, and the other being a lengthening of the food grooves on the theca. One at least of these trends—the lengthening of the thecal food grooves—apparently happened independently in the two groups into which the class is divided: Diploporita and Rhombifera[95]. This taxonomic division is based on the structure of the special thecal pores, which are called *diplopores* and *rhombopores*.

Diplopores

As their name suggests, these normally pierce the theca in pairs, though single pores (haplopores) also occur in some forms. In surface view (Fig. 18*a*) a single diplopore shows an oval depression with the two pores near each end. Each pore passes right through the plate, and the opening on the inside is sometimes bordered by tiny projections from the plate into the interior of the body. No cystoid has ever been described with remains of any of its soft parts, so any reconstruction must be made solely by comparison with living members of other classes of echinoderms. On this basis, one possible explanation is that each diplopore bore a thin-walled bulb of tissue (Fig. 18*c*) for respiration, rather like the papulae of modern asteroids (p. 38), but instead of a single pore the diplopores had an 'entrance' and 'exit' to ensure circulation of fluid in the organ. On the inside of the test a similar thin-walled bulb would have protruded into the coelom. So gaseous exchange could take place between sea-water and respiratory fluid across the wall of the external bulb, and between respiratory fluid and coelomic fluid across the wall of the internal bulb. The function of the inward-directed projections was most likely to support the inner soft bulb in such a way that it did not sag against the inner face of the theca, which would be detrimental to respiratory efficiency.

Rhombopores and pectinirhombs

In the Rhombifera the structure of the pore systems is markedly different from that of the diplopores. Here, instead

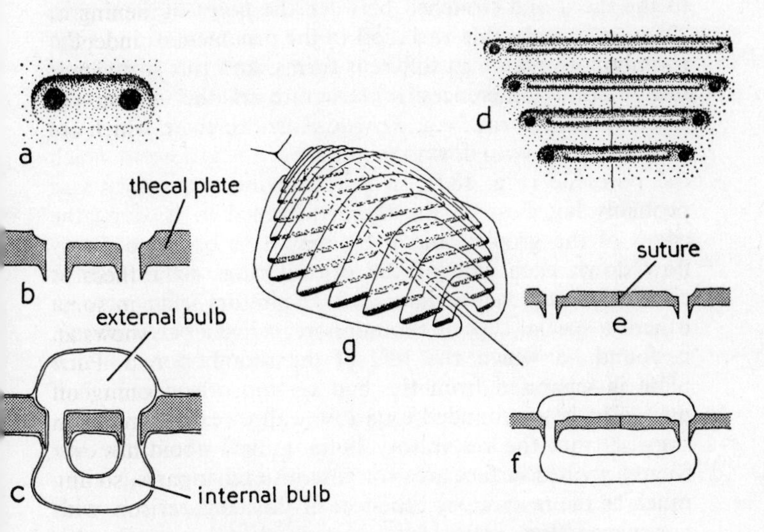

Fig. 18 STRUCTURE OF CYSTOID PORE SYSTEMS

a Surface view of a diplopore.
b Section across diplopore.
c Tentative reconstruction of a diplopore in section. The external and internal thin-walled sacs may have had walls thrown into folds.
d Surface view of part of a pore-rhomb, based on that of *Echinosphaerites*. The grooves and ridges traverse an inter-plate suture.
e Section of *d*, along one groove.
f Tentative reconstruction of part of a pore-rhomb, in section. As in *c*, the walls may have been folded.
g Perspective view of a tentative reconstruction of a pectini-rhomb. Each pair of slits probably gave rise to an arch-shaped, thin-walled sac.

of the pore-pairs being dotted about all over the theca, they are aggregated into groups, each group lying across an interplate suture. The distance between the pores of a pair is greatly increased in the centre of each group, though it diminishes at either end (Fig. 18*d*), forming a rhomb-shaped area. A possible reconstruction of this (Fig. 18*g*) shows a series of thin-walled bulbs like short gill lamellae, lying close to the theca and stretched between the pores of each pair. There is considerable variation in the ornamentation of the rhombopore region in different forms, and this must have meant some differences in structure of the respiratory sacs. In some forms, e.g. *Echinosphaerites*, there is a ridge (or sometimes two) down the middle of the groove in which the pores lie (Fig. 18*d*); in this case the respiratory sacs probably lay close to the theca, attached to it along the edges of the groove, and there may have been a one-way flow down each side of the central ridge or ridges for circulation and hence increased respiratory efficiency. In others, a special type of rhombopore, called a *pectinirhomb*, is found, in which the half of the rhombopore on one plate is separated from the half on the other; each half may even be surrounded by a low wall of calcite, and this suggests that the respiratory bulbs looped above the test to give greater surface area for gaseous exchange (Fig. 18*g*), much as the respiratory tube-feet of the irregular echinoids are shaped (Fig. 14*j*).

A section of a pore-rhomb, such as that of *Lepadocrinus*, shows that in some rhombiferans the depressions between the pores of each pair were deep grooves which did not open to the interior of the theca along their entire length but were rather in the form of elongated funnels, leading down to the pore. The texture of the infilling of these grooves suggests that each one may have held a thin-walled sac whose lumen and part of the space between the sac and the thecal plates became filled in and secondarily calcified in a manner different from that of the sac wall. If this is so, it indicates a remarkable parallel with the groove underlying many tube-feet of recent echinoderms, which is lined by a sheath of connective tissue and coelomic epithelium continuous with that forming the wall of the rest of the tube-foot/ampulla system. The parallel between the structure of a pectinirhomb and that of a group of respiratory tube-feet of an irregular echinoid is particularly noticeable, though the respiratory lamellae of a pectinirhomb had to be packed closer together, as Bather[89] showed, to make room for the gut inside the body.

It must be said that opinions have differed, not only on the interpretation of the structure of cystoid pore-systems[90], but also on their structure in the fossils. Some authors[88] would have it that the entire theca of cystoids was covered by an additional thin calcite film (the 'epistereom'), and some even say that there was another on the inside as well (the 'hypostereom')[1]. All workers seem to agree that the function of the pore-system was most likely respiratory, but how gaseous exchange is supposed to have occurred across calcite layers is not explained, unless they are seen as exceedingly thin sheets, as in the hydrospires of blastoids (p.144): apart from this, it is in any case not rare for fossils to gain an extra coating of calcite, secondarily deposited, during the process of fossilization (p.94). Indeed, in many specimens, particularly of *Echinosphaerites*, it is possible to see that the extra covering of calcite, both inside and out, is continuous with that filling in the pores and grooves of the system, so it is most likely a secondary deposit. But unfortunately this is not the whole story; a very puzzling situation occurs in *Heliocrinus* (Upper Ordovician) which rather puts doubt on the interpretations suggested above. In this rhombiferan one can see without any doubt that the channel between the individual pores of at least some pairs was covered over by a thick layer of calcite continuous with that of the theca, and as far as can be seen no other pores emerge from the covered canal. The material I have been able to examine of *Heliocrinus* and other cystoids showing canal systems which were definitely closed over is unfortunately somewhat fragmentary, and it is not possible to see for certain whether the closed condition is common to all canals of all pore-rhombs, but one gets the impression that only the longest canals, that is, those in the centre of each rhomb, are closed. If this is so, then it might be that the others had the usual thin-walled respiratory sacs while these possessed flow receptors as balance organs, in which the differential flow of fluid in the organs against sensory nerve endings served the important function of maintaining an upright position. What is interesting in this connexion is that some cystoids (but not, as far as I know, those showing definitely closed canals) have a plate arrangement at the basal tip suggesting a sucker-like structure (for instance, *Aristocystites*[96]). If these had temporary fixation only, it would be important for the animal to keep upright, and it is not beyond the bounds of possibility that selection modified an existing system for the purpose. This is speculation, but at least it fits all the known facts.

The evolution of the cystoids

Though there are unreliable accounts of some cystoids from the Cambrian, it is probably true to say that the Ordovician saw the establishment and main deployment of the group. There are a few forms in the Silurian and Devonian, but later than this the group dies out. Since the stratigraphical horizon of most of the specimens has been inadequately recorded, it is not possible to give an accurate time-sequence to the main trends in the Ordovician. One of the simplest forms to appear in this period is the diploporite *Aristocystites* (Fig. 19*b*). This is flask-shaped, with a theca composed of many irregularly arranged plates pierced by haplopores and diplopores. It is assumed that arm-like extensions, the *brachioles*, bearing food grooves arose from the area round the mouth, but they have never been found intact. This rather simple form of the cystoid condition continues into the Silurian, where the somewhat similar *Holocystites* occurs, and here some of the specimens do show perioral brachioles. Contemporaneous with this series is an equivalent one in the Rhombifera: *Echinosphaerites* (Fig. 19*g*) occurs at roughly the same time as *Aristocystites*, and it too had a haphazard arrangement of thecal plates, and the food grooves were restricted to the perioral brachioles. This simple condition is also true of the later rhombiferan *Callocystites* (Silurian), which Bather[1] suggests may have evolved from *Echinosphaerites*-like forms. It is interesting to recall that *Echinosphaerites*, found in rocks near the Baltic as round balls filled with radiating crystals of calcite, were included by Linnaeus in his Mineral Kingdom under the name *aetites*, and were known to the older Scandinavian naturalists as 'crystal apples' until 1772, when Gyllenhaul discovered their true nature.

Returning to the Ordovician, we find there are members from both orders whose food grooves extend over the theca in various ways, and in some, at least, the grooves are bridged by movable plates, rather like the food grooves of living crinoids. Thus, in the Diploporita we can trace a series (though not necessarily a phyletic one) starting with such forms as *Eucystis* (Fig. 19*c*) in which the brachioles, usually five, arose from only a little way away from the mouth; through *Glyptosphaerites* (Fig. 19*d*) and *Proteocystis*, in which the grooves are longer and the brachioles are increased in number and arise from branches of the five main grooves; to *Fungocystis* (Fig. 19*e*) and *Protocrinus*, where the food grooves extend to the base of the theca. A modification of this last pattern is seen in such forms as *Proteroblastus* (Fig. 19*f*), in which all the diplopores are aggregated close to the food grooves, possibly to derive benefit from the current in that area for respiratory purposes.

In the Rhombifera a similar trend can be traced from *Echinosphaerites*-like forms, in which there are effectively no grooves on the theca, the brachioles arising directly from the perioral region; through forms like *Cystoblastus* (Fig. 19*h*), in which the food grooves and associated brachioles are borne only on the adoral surface; to *Lepadocrinus* (Fig. 19*i*), in which the grooves and brachioles extend nearly to the point of insertion of the stem.

The trend towards an orderly arrangement of thecal

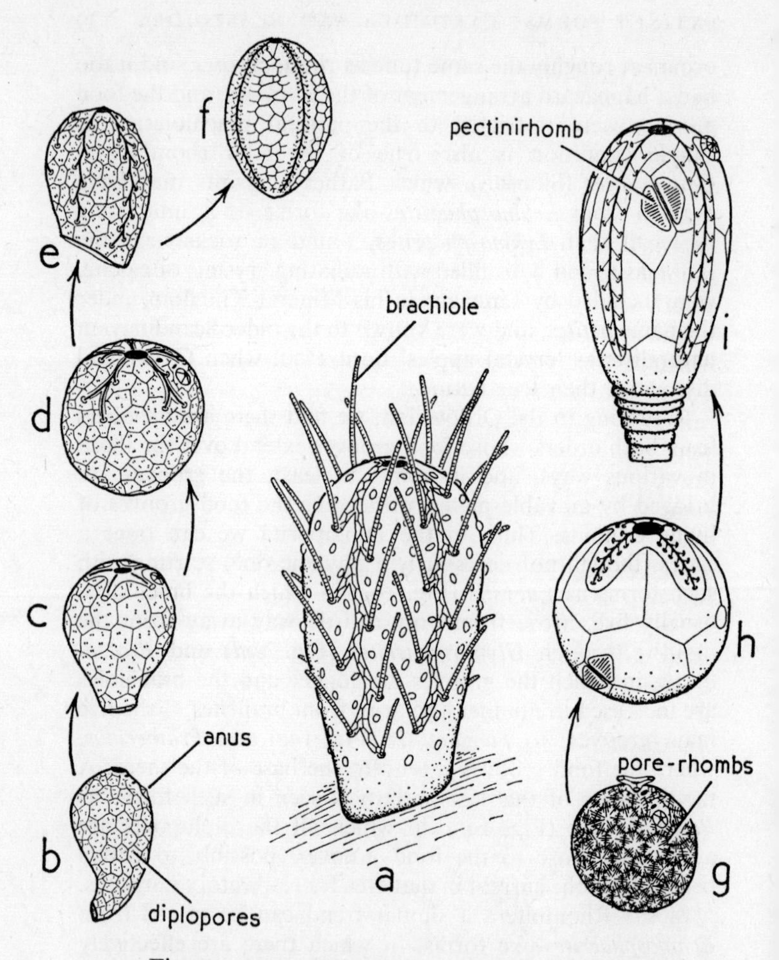

Fig. 19 RANGE OF BODY FORM IN THE CYSTOIDS

a Diagrammatic reconstruction of a typical cystoid, based on the diploporite *Fungocystis* (see *e*), with cover-plates closed over the food grooves. There is no evidence indicating the length of the brachioles.

b to *f*, evolutionary trends in the Diplopoirta, showing the progressive lengthening of the food grooves over the theca.

b *Aristocystites* (Ord.).

plates and an associated reduction in the number of plates seems to have happened only in the Rhombifera; it is a trend associated with the reduction in the number of 'respiratory' areas from a large number of pore-rhombs across most of the interplate sutures, as in forms like *Echinosphaerites*, to pectinirhombs in restricted places only, as in *Lepadocrinus*. In these latter forms the plates become arranged in whorls with usually five plates to each whorl, very much like the theca of crinoids.

The cystoids were, then, flask-shaped animals, some of which were attached to the substratum by a sucker and others by a stem, which held out brachioles to catch the rain of food dropping to the ocean floor. In some the plates were numerous and haphazardly arranged, while in others fewer plates were arranged mainly in whorls of five; in some the brachioles arose from the region of the mouth, while in others they arose from the sides of food grooves on the theca; in some, what appear to have been respiratory organs were scattered over the entire theca, while in others these structures were restricted to special areas. It is on this latter difference that two orders are recognized within the Cystoidea.

The Blastoidea

This group shows so many similarities of thecal and brachiole arrangement with the later cystoids, that is, those

---

c *Eucystis* (Ord.).
d *Glyptosphaerites* (Ord.).
e *Fungocystis* (Ord.).
f *Proteroblastus*, in which the diplopores are restricted to the region of the food grooves.

g to *i*, equivalent trends in the Rhombifera, showing also the restriction of the pore-rhombs to a few areas only.
g *Echinosphaerites* (Ord.–Sil.).
h *Cystoblastus* (Ord.), with two pectinirhombs.
i *Lepadocrinus* (Sil.), with three pectinirhombs, of which one is shown. The arrows in the diagram do not necessarily imply phyletic lines.

that have attained pentamerous symmetry, that some authorities[93, 95] favour placing them as an order of the Cystoidea. On the other hand, the arrangement of the structures which are generally agreed to have been at least in part respiratory is so different from that of the pore-rhombs that it seems desirable to separate the group as a class, as is more usual[7]. The theca (Fig. 21*a*) is bud-shaped, with five food grooves radiating from the upwardly directed mouth; in life numerous brachioles arose from the sides of its food grooves, and sometimes these have been preserved *in situ*. The basic plate-arrangement of the theca is simpler than the simplest arrangement found in the cystoids: three whorls of plates make up the whole theca, except for the food grooves. Each whorl consists of five plates: round the mouth is a ring of deltoids, between and below them a ring of radials, usually the largest plates of the theca and notched orally to receive the food grooves, and below these are the basals, which in later blastoids become secondarily reduced by fusion to three plates, two large (each formed by the fusion of two plates) and one small. The stem is attached to the basals and breaks up into roots at its lower end for insertion into the substratum. In the great majority of blastoids the slight irregularity in the basal whorl of the theca and the presence of the anus in one inter-radius is the only departure from radial symmetry, but in some there is a much greater departure from radiality: one food groove may be reduced (as in *Eleutherocrinus*) or modified in structure (as in *Zygocrinus*), making the theca bilateral.

The most intriguing feature of the blastoids is the structure of the food grooves and neighbouring organs. Some specimens are so well preserved that it is possible to make transverse sections of the theca by grinding, so that a considerable amount of the anatomy of these regions, even of the soft parts, can be made out. One can see that the structure was more complicated than just a ciliated food groove: there was at least one canal running below each groove, embedded in its plates, the five canals of this system connecting up with a circum-oral ring canal. Also, the system of respiratory organs, which will be described later, is closely associated with the food grooves, so it is evident that soft structures other than the epithelium of the food grooves have become radially placed and have assumed pentamery. For this reason we may use the term *ambulacrum* for each groove and its associated structures in each radius. A blastoid ambulacrum, as mentioned above, lies in a notch in the oral side of a radial plate. This notch is filled mainly by a single elongated plate, the *lancet* (Fig. 20*a*),

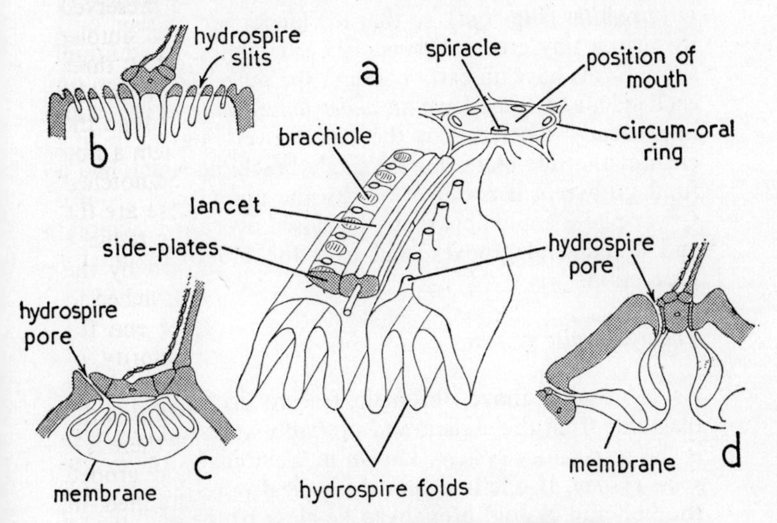

Fig. 20 STRUCTURE OF BLASTOID AMBULACRA

a Schematic perspective diagram of part of one ambulacrum of a typical blastoid, based on that of *Pentremites*. The side-plates of the right side have been removed.

b T.S. ambulacrum of *Codaster*, showing hydrospire slits.

c T.S. ambulacrum of *Pentremites*, showing ingrowth of hydrospire folds. The hydrospire pores alternate, so are shown on the left side only.

d T.S. ambulacrum of *Orbitremites derbiensis*, showing single hydrospire folds. The side-plates overlie the lancet.

In *b*, *c* and *d* a brachiole and the cover-plates to the food grooves are shown on the right side only.

forming the floor of the ambulacral groove. This plate extends beyond the oral end of the radial containing it, and so comes to lie against the two adjacent deltoids, which are excavated to contain it. A longitudinal canal runs within this plate, passing inwards at the oral end through the interdeltoid suture and then branching before joining the circum-oral ring canal. In addition to the lancet there is normally a column of *side-plates* between its lateral margins and the notch in the radial, as in *Pentremites* (Fig. 20*c*); or the side-plates may come to lie on top of the lancet, as in *Orbitremites* (Fig. 19*d*), so that the lancet can be seen only through a tiny groove between the side-plates. In addition, some forms have an extra column, the *outer side-plates*, on each side, and some have an *under lancet-plate* in the mid-line usually shorter than the lancet overlying it. In life, each of the side-plates bore a single brachiole which had a food groove in it continuous with the ambulacral groove. Cover-plates could apparently be closed over the brachiolar and ambulacral grooves, and over the mouth region for protection.

The hydrospire system

As mentioned above, the main feature distinguishing the blastoids from the earlier, and probably ancestral, cystoids is the respiratory system, known in blastoids as the *hydrospire system*. If one imagines the typical pore-rhombs of a rhombiferan cystoid brought to lie close to the ambulacral grooves so that they traverse the radio-deltoid sutures in each interambulacral area, one can see how the basic situation in the blastoids possibly arose, a situation which is foreshadowed in the cystoids themselves by the diploporite *Proteroblastus* (the name itself suggests this), which had its respiratory organs alongside the grooves. But instead of being mere indentations in the surface bearing external soft parts in contact with the sea-water, the blastoid system consisted of slits leading right into the interior of the theca, and from these slits on the inside hung folds of tissue like the folds of a curtain (Fig. 20*a*). The reason why we can say this with some certainty is because the tissue of the interior folds was apparently supported by a calcareous skeleton derived from the lancet and deltoid-plates. The cavity of the theca is often filled by a single crystal of calcite, and if a transverse section of a blastoid is prepared by grinding, one can see a faint outline of the hydrospire folds on the ground surface and can follow them a good way down from the plane of the section into what was the body cavity. Since the skeleton supporting these folds could have had the usual fenestrated structure of echinoderm plates (Fig. 12*j*) it is possible that a considerable proportion of the area of the folds was soft tissue, across which exchange of gases took place. In sections of some blastoids one can see other structures strikingly similar in appearance to the hydrospire folds, that is, also looking like thin membranes. One of these appears to lie fairly close to the inside wall of the theca, even attached to it in places, and to enclose the hydrospire folds in isolation from the rest of the coelom. In forms like *Pentremites* (Fig. 20*c*) this membrane isolates the hydrospires of one ambulacrum in a single sac, which is in consequence radially placed; but in *Orbitremites*, for instance (Fig. 20*d*), the hydrospire of one side of an ambulacrum is isolated with its neighbour from the adjacent ambulacrum, thus lying in an inter-radially situated sac. This situation is, of course, quite unlike anything found in the living classes; but then no living echinoderm has hydrospires. The nearest we get to it is the membrane separating the inner and outer coeloms in crinoids (p.26).

One can trace a series, possibly phyletic in view of the time-relations, based on the structure of the hydrospire system. Early blastoids, such as *Codaster*, Silurian (Figs. 20*b*, 21*c*), show a situation closest to that in the cystoids: a series of parallel splits runs at right angles across the radio-deltoid suture in each interambulacrum, each slit leading inwards to a single fold of soft tissue. Later, in forms such as *Pentremites* (Carboniferous) the five or so folds become pushed into the theca and open by a longitudinal row of pores, the *hydrospire pores*, usually placed between the side-plates (Fig. 20*c*). In addition, each interambulacrum has a single pore, the *spiracle*, which connects with the hydrospire cavities on adjacent sides of neighbouring ambulacra (Fig. 20*a*). The pores probably provided an entrance to the hydrospire system for the sea-water, the folds a respiratory surface for the exchange of gases between sea-water and coelomic fluid, and the spiracles exhalent pores. In some forms with this one-way system the number of folds is reduced to one on each side, as in *Orbitremites derbiensis*, while in others there may be up to nine folds on each side, as in *Neoschisma*.

The evolution of the blastoids

It seems to be generally agreed[1, 92] that the earliest known blastoid was *Blastoidocrinus*, Middle Ordovician (Fig. 21*b*). This is occasionally classified with the cystoids, because its respiratory organs, though parallel with the food grooves, did not pierce the thecal plates in the manner of the hydrospires but arose from grooves in the plates like the pore-rhombs. Another cystoid feature is that there were many aboral thecal plates instead of whorls of five. But on the

---

Fig. 21 EVOLUTION AND RANGE OF BODY FORM IN THE BLASTOIDS

a Reconstruction of a typical blastoid, such as *Orophocrinus* (Carb.).

b to *f*, the main evolutionary trends within the blastoids.

b *Blastoidocrinus* (Ord.), showing irregular thecal plate arrangement and hydrospire slits.

c *Codaster* (Sil.), with ambulacra still confined to oral region.

d *Orbitremites*, and *e* *Pentremites* (Carb.), in which ambulacra extend nearly to stem insertion. Entry to hydrospires by pores.

f *Pterotoblastus* (Perm.), with ambulacra borne on arm-like processes.

g and *h*, bilaterally symmetrical blastoids, each with one ambulacrum modified.

g *Eleutherocrinus* (Dev.). i) side view, ii) oral view.

h *Zygocrinus* (Carb.). i) side view, ii) oral view.

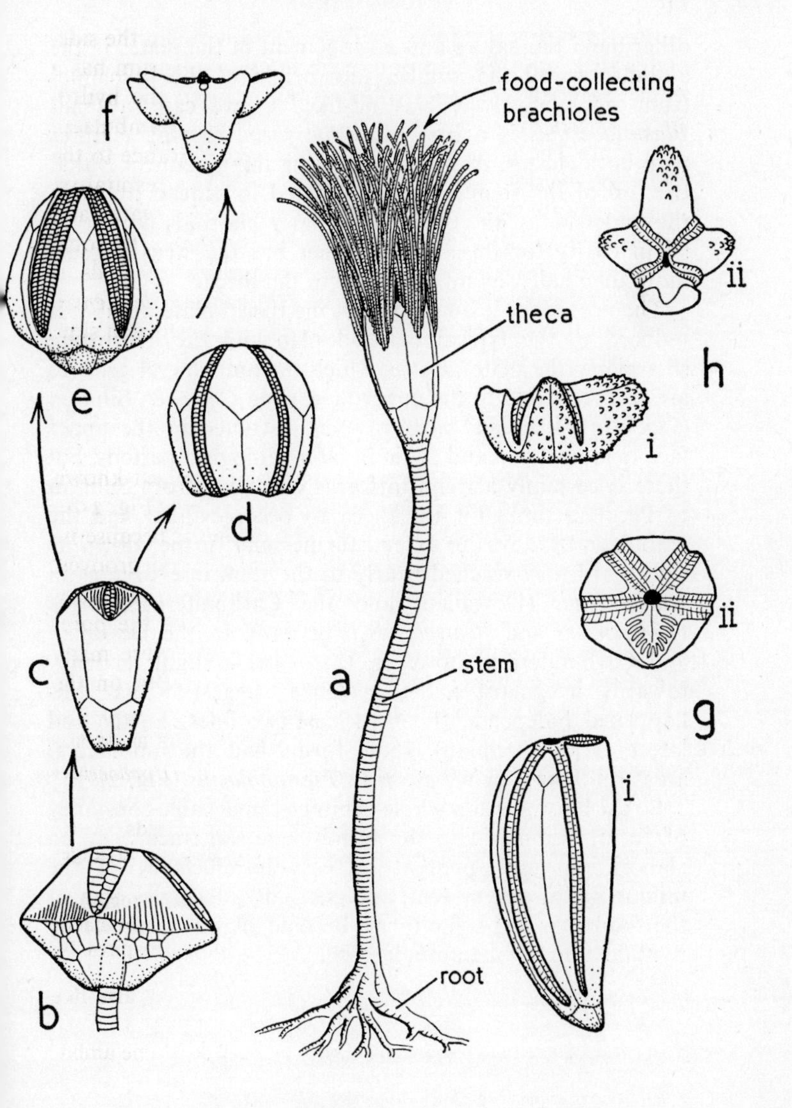

food-collecting
brachioles
theca
stem
root
a
b
c
d
e
f
g
ii
i
h
ii
i

other hand the nature and arrangement of the plates in the ambulacra and the number and origin of the brachioles from side-plates was blastoid-like, so we can say that *Blastoidocrinus* is a true *transition form*, sharing features with both classes. Another interesting thing about it is that the top of the stem is accommodated in a deep cavity in the underside of the theca; some later blastoids do have a slight cavity for this, but no other has the stem inserted more than half-way up the inside of the theca.

The next stage is one in which the thecal plate number is reduced to the typical blastoid plan and the hydrospire slits pierce the plates, but in which the ambulacral grooves are still restricted to the upper face, as in *Codaster*, Silurian (Fig. 21*c*). Blastoids with grooves restricted to the upper face turn up again and again in later geological periods, but there is certainly a trend in forms occurring from Silurian to Permian times for the theca to become taller and the ambulacral grooves to extend further and further down its sides until they reached nearly to the stem insertion, as in *Nucleocrinus* (Devonian) and the Carboniferous forms *Orbitremites* and *Pentremites* (Fig. 21*e*). In the Devonian and Carboniferous, however, there was a slight tendency towards irregularity; the evidence suggests[92] that this happened independently in at least two lines. Lastly, and latest of all (Permian), some forms had the ambulacral areas on flanges or wings, as in *Pterotoblastus* (Fig. 21*f*).

So the blastoids as a whole exhibit a remarkable constancy of form. Summarizing the trends, one can trace, first, an elaboration and apparent increase in efficiency of the unique hydrospire system, and, secondly, a lengthening of the food grooves; some forms became slightly irregular by modification of one ambulacrum.

12

EDRIOASTEROIDEA, HETEROSTELEA AND LESSER-KNOWN EXTINCT GROUPS

The Edrioasteroidea

The usual textbook figure of a generalized edrioasteroid looks rather like a starfish hugging a piece of concrete (see, for instance, Fig. 22*g*). The 'starfish' is actually five ambulacral grooves, and the 'concrete' the interambulacral areas. The group ranges from the Mid-Cambrian to Lower Carboniferous, and is usually considered of particular interest because from the point of view of their time-relations and certain morphological features they may have been close to the ancestor of all the Eleutherozoa, and particularly the asteroids. Bather, whose *Studies on the Edrioasteroidea*[98] has advanced our knowledge of this important group to a great extent, recognizes four families: the AGELACRINIDAE appear to be primitive, with a thin flexible test and no ambulacral pores; the EDRIOASTERIDAE have ambulacra that pass on to the aboral surface; the CYATHOCYSTIDAE have their interambulacral plates fused to form deltoids; and the STEGANOBLASTIDAE are superficially blastoid-like, with a rigid theca of thick plates.

The group is usually referred to the Pelmatozoa, but though some of them, such as *Pyrgocystis* (Fig. 22*f*), were provided with a fairly typical pelmatozoan stem, others, such as *Edrioaster* (Fig. 22*g*) and *Dinocystis*, two of the best-known forms, are stemless, have a concave undersurface and are never found in rocks suggesting a hard

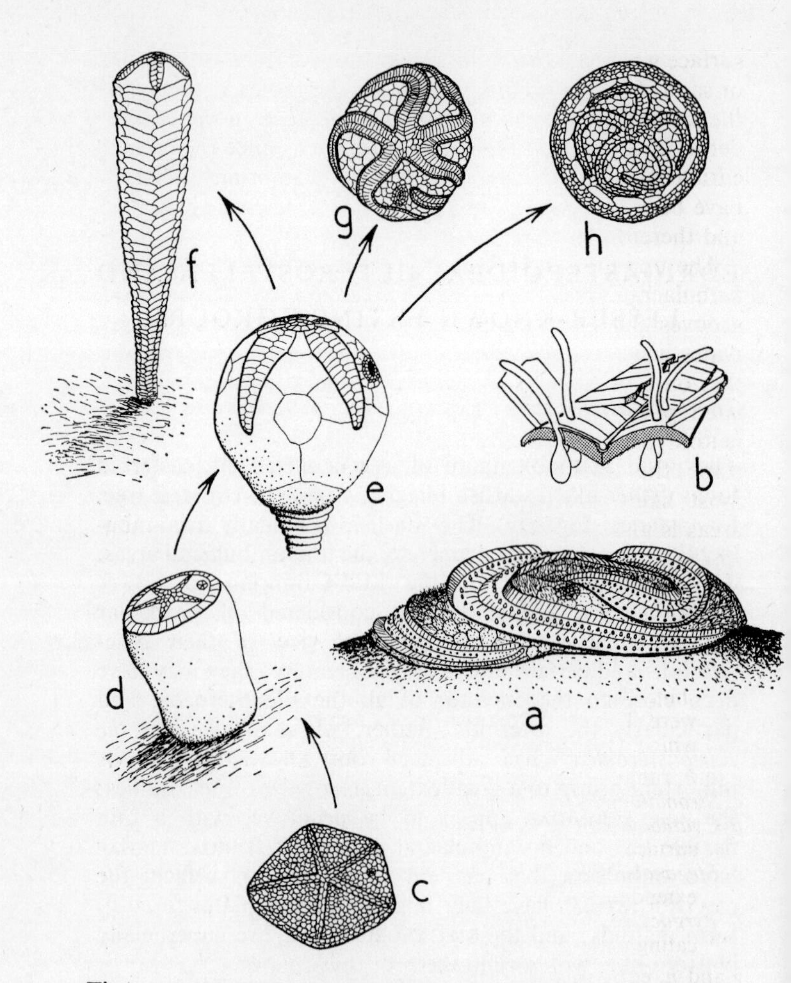

Fig. 22 STRUCTURE AND RANGE OF BODY FORM IN THE EDRIOASTEROIDS

a Reconstruction of a typical edrioasteroid, with ambulacral cover-plates open but tube-feet retracted, based on *Edrioaster*.

b Tentative reconstruction of part of one ambulacrum. The first two cover-plates on each side are open, the rest almost surface when laid down but rather those derived from shales or sand. In one or more lines within the group a reversal of the evolutionary trend towards free life, seen in the echinoderms as a whole, seems to have occurred, since the earliest edrioasteroid (*Stromatocystis*, Middle Cambrian) seems to have been free, while some of the later ones were stemmed and therefore probably fixed.

The upper (oral) surface consists in general of five interambulacral areas separated by the five ambulacral food grooves. The interambulacra consist of a mass of irregular plates which are sometimes imbricate (overlapping, like the tiles on a roof) as in *Agelacrinus*, or abutting (like a tesselated pavement) as in *Edrioaster*, or fused into one large deltoid plate per interambulacrum as in *Cyathocystis*. In some the interambulacral plates show tubercles, which most likely bore spines. In one of the interambulacral areas is an anus borne on a spire consisting, as in cystoids (p.133), of a ring of five plates, and, generally close to the anus, is a hydropore. The food grooves radiate out from a central mouth, which is surrounded by a strengthening

---

shut. There is no evidence to show how long the tube-feet were. This reconstruction does not apply to the agelacrinids, which apparently had no ampullae.

c to *h*, range of body form. All drawn with cover-plates closed.
c *Stromatocystis* (Camb.), probably the simplest edrioasteroid.
d *Cyathocystis* (Ord.), with straight ambulacra confined to oral surface.
e *Steganoblastus* (Ord.), a stemmed form, with ambulacra extending further over the theca.
f *Pyrgocystis* (Ord.–Sil.), with sturdy stem enclosed by imbricating plates; ambulacra restricted to oral surface.

g and *h*, edrioasteroids with curved ambulacra.
g *Edrioaster* (Ord.–Dev.), with one ambulacrum curving in a direction opposite to that of the others.
h *Agelacrinus* (Dev.), with two ambulacra curving the other way.

The arrows do not necessarily imply phyletic lines, though time-relations and comparative structure are consistent with this being a rough family tree of the edrioasteroids.

frame consisting of five radial and five interradial pieces. In some forms, chiefly those which seem to be primitive, the food grooves are straight, while in others they are curved; the curving may be in the same direction, or, more often, one or two rays may curve in the opposite way, to surround the anus and hydropore (Fig. 22*g*, *h*). Both grooves and mouth region were covered by roofing plates, some of which remain in well-preserved specimens; those over the grooves were probably movable, while those over the peristome were fixed. The food grooves are floored by two columns of alternating plates; between the plates of each column are pores leading down to the interior of the theca. It is possible to say with some certainty that these pores provided a connexion between the tubes of the water vascular system outside and the ampullae of tube-feet inside, because in one species particularly, *Edrioaster buchianus*, tiny grooves have been seen in the flooring plates round the mouth, probably representing the seat of a circum-oral water ring, and others running perradially down the centre of each groove, probably for radial water vessels. The water ring was connected to the hydropore canal through a pore in the mouth frame. This is a palaeontological windfall, for it tells us that the edrioasteroids had an 'open' ambulacral system, like the asteroids and crinoids (p.44) among modern echinoderms.

Nothing can be said with any certainty about the nature of the tube-feet. The Agelacrinidae had no pores between the groove plates and no hydropore, and these facts may mean that they had no ampullae to their tube-feet. They may even have lacked radial water vessels and tube-feet altogether (certainly no grooves for the vessels have been recognized), though the configuration of the cover-plates, according to Bather[98], seems consistent with the extrusion of podia between them—podia, that is, without the power of protraction and retraction. If this is so, then it is tempting to suggest that this family, which includes the earliest-known edrioasteroid, possessed passive tube-feet whose function was purely respiratory; later forms, of the other three families, had the additional advantage of prehensile tube-feet, actuated by the hydraulic ampulla system, perhaps an innovation which could be used to advantage when feeding. One imagines these later forms feeding by opening the cover-plates (Fig. 22*a*, *b*), extending their tube-feet to increase the catchment area, and possibly even bending their tube-feet in a crinoid-like manner (p.110) to help food into the groove. On retraction, the tube-feet were probably brought into the groove and protected by the closed cover-plates.

One feature, in forms like *Edrioaster* and *Dinocystis*, is of interest. This is the structure of the concave under (aboral) surface, which looks as if it possessed a tough membrane stretched across a central area bordered by a rigid skeletal frame; the membrane was most probably a sucker for temporary fixation. Nothing is left of the muscle system which might have operated this mechanism; neither have any muscle scars been described from the interior of the theca. The animal probably moved by rhythmic movements of the flexible theca. Certainly, some slightly irregular forms have been found which sag to one side, in such a position in relation to currents, as indicated by the orientation of brachiopods in the same beds, that they probably had the power to move, limpet-like, so that the anus was always to the leeward side of the mouth[98]. For an animal dependent on food settling in the grooves this would be important, and the suggestion is attractive for this reason. The only other echinoderms which look as if they may have had a basal sucker are some cystoids (p.138).

The evolution of the edrioasteroids

The origin of the group, as Bather[98] has pointed out, must have been during or before the Lower Cambrian, so that fossil evidence is lacking. From comparative anatomy, and from time-relations, the most probable origin is from a primitive crinoid, possibly a form like *Stephanocrinus* or *Cyathocrinus*, by losing the arms and pinnules and by a general flattening of the theca. The structure of the crinoid tegmen is certainly consistent with this view.

Of the evolution within the class little can be said with certainty. The earliest form is the agelacrinid *Stromatocystis*, Cambrian (Fig. 22*c*). Other agelacrinids are found in all subsequent periods up to the Carboniferous; e.g. *Pyrgocystis*, Ordovician (Fig. 22*f*); *Hemicystis*, Ordovician to Silurian; *Agelacrinus*, Devonian (Fig. 22*h*); and *Lepidodiscus*, Ordovician to Carboniferous, very similar to *Agelacrinus*. It seems likely that the Cyathocystidae branched from the agelacrinid stock fairly early[1], and the Steganoblastidae (Fig. 22*e*) might have done likewise. So possibly did the Edrioasteridae, because although this family contains the well-known forms with curved ambulacra (e.g. Fig. 22*g*), and for this reason could have evolved from *Agelacrinus*-like forms, some early edrioasteroids such as *Aesiocystis* (Ordovician) had straight ambulacra and a plate arrangement consistent with their having evolved from a *Stromatocystis*-like agelacrinid[1]. Of course, the taxonomic arrangement used here may not reflect phyletic lines, but whatever the true relationships are, we seem to have a good example of 'explosive evolution' in the Ordovician, after the basic edrioasteroid plan had been evolved in the Cambrian.

As Bather[98] and Neumayr have pointed out, there are interesting similarities between the plating of edrioasteroids and asteroids which attract attention to the possible connexion between these two groups. First, there are the obvious similarities in the flooring plates of the ambulacra and the fact that both groups have or had open ambulacra; secondly, the edrioasteroid interambulacral plates may well be homologous with the virgalia of early asteroids; and thirdly, the elements of the early asteroid mouth frame can be derived from the edrioasteroid pattern. As to time-relations, no asteroid is known before the Lower Ordovician, but the edrioasteroids are known from Mid-Cambrian, so they certainly have good claim to be ancestral to the asteroids. What must be emphasized is the fact that the edrioasteroid and asteroid lines must have evolved internal ampullae independently, because the earliest asteroids lacked pores to the interior (p. 48).

The Heterostelea (= *Carpoidea* or *Amphoridea*)

Apart from the Machaeridea, which may not be complete animals anyway, the Heterostelea is the only group of the echinoderms which does not show radial symmetry in any of its members. As in the case of machaerids the chief reason for including them in the phylum is the crystal structure of the thecal plates; but rather stronger additional grounds exist here, because some specimens can be seen to possess brachioles with the usual echinoderm food grooves running their length on one side (e.g. *Dendrocystis*, Fig. 23*c*). Bather[100] regards them as a somewhat specialized early offshoot from the main deployment of the phylum, an offshoot which may represent the brief appearance of forms having a body plan resembling the hypothetical *dipleurula* (p. 175). His idea is that the bilateral heterostele metamorphosed from the bilateral dipleurula and did not pass through a radially symmetrical stage, and for this reason, as well as their time of appearance (Middle Cambrian), he thinks they could be a 'stem group' to the phylum which quite early on gave rise to the radiate forms. But this must have happened during the early part of the Cambrian period or before, and there are no fossils to back up the hypothesis.

In general terms, the heterosteles all possess a flat theca the plates of which differ on the two sides. There is typically a stem formed of several columns of imbricating or abutting plates (which Bather[104] suggested might be the source of the machaerids, see p. 164) and there may be one or more spines or brachioles. Some forms (particularly *Gyrocystis*[103]) have a plate structure in the oral region which suggests that the mouth was surrounded by a ring of soft tentacles the whole set of which could be withdrawn into the body and covered by an operculum. Let us examine *Gyrocystis* and its allies first, because not only are they among the best known heterosteles but they are also the earliest (Middle Cambrian).

The theca (Fig. 23*a*, *b*) consists of a ring of about twelve large marginal plates and an upper and lower sheet of plates, the upper, which is convex, consisting of very small loosely joined plates and the lower (concave) of slightly larger and close-fitting ones. At one end the ring of marginal plates is pierced by two holes of unequal size, the larger of the two being covered by an apparently movable plate, the *operculum*. Inside the larger hole is a set of ossicles cutting off a small cavity, the *stomodeal pouch*, just inside the hole. At the opposite end is the stem, consisting of between four and six longitudinal columns of plates and gradually tapering to a point. Spencer[103] has put forward what seems to be the most sensible and most widely accepted theory as to the mode of life of this form and its close relative *Trochocystis* (Middle Cambrian). He thinks that the stomodeal pouch housed a set of protrusible tentacles which acted as a lophophore for catching food; he points out that many specimens are found with the stem preserved at right angles to the plane of the theca, as though temporarily inserted into the substratum; he also makes the interesting point that there is a tremendous variation of body shape in the many specimens which have been found. These facts suggest that the animal remained semi-sessile on the sea-bottom and searched the surrounding area with its tentacles, moving its body here and there to do so and changing shape to reach all the area available to it, much as some of the dendrochirote holothuroids wedge themselves in a rock crevice and elongate and twist the body to search the surrounding region with the oral tube-feet.

One feature which has puzzled investigators is the presence in both *Gyrocystis* and *Trochocystis* of grooves radiating from the smaller hole (on Spencer's theory the anus). These were naturally thought to be food grooves, and their presence led some authorities[101, 102] to regard the smaller hole as the mouth and the larger as the anus. An explanation for the large size of the anus was that the Heterostelea had

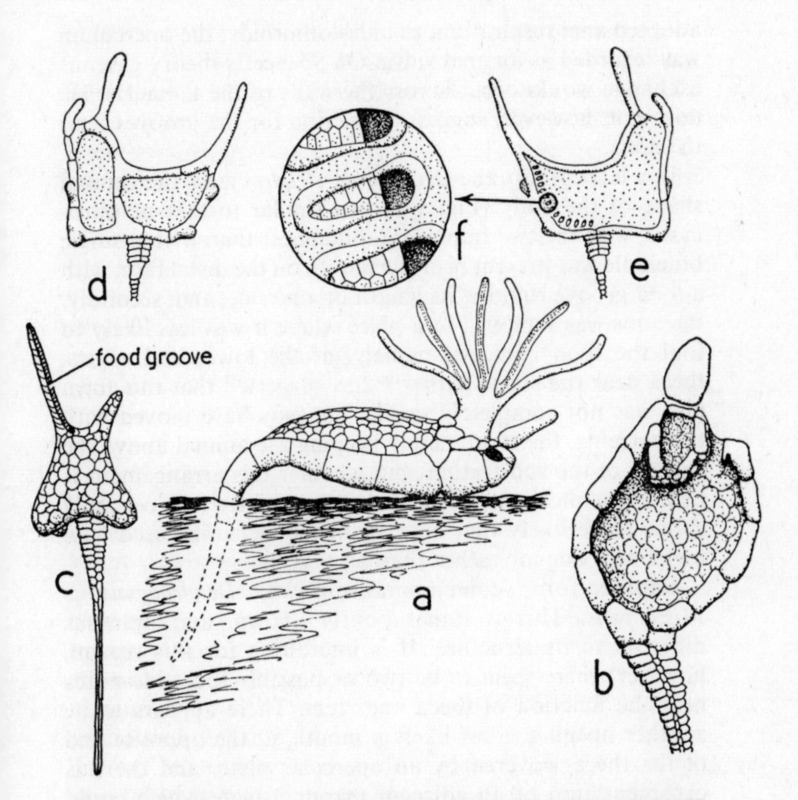

Fig. 23 MORPHOLOGY AND RANGE OF BODY FORM IN THE HETEROSTELEA

a Tentative reconstruction of *Gyrocystis* in its feeding position.
b The fossilized remains of *Gyrocystis* (M. Camb.) seen from above, with the operculum drawn in the open position.
c *Dendrocystis* (Ord.), with single brachiole.

d to *f*, *Cothurnocystis* (U. Ord.).
d Underside.
e Upper surface, showing column of special pores.
f Part of the column of pores enlarged, showing cover-plates.

adopted anal respiration, as in holothuroids; the operculum was regarded as an anal valve. On Spencer's theory gaseous exchange would occur across the walls of the tentacles. He does not, however, suggest a function for the grooves near the anus.

In a later group, the Ordovician *Dendrocystis*, the general shape of the body (Fig. 23*c*) was similar to that of *Gyrocystis*, with the two important differences that, first, a single brachiole was present near the mouth on the distal face, with a food groove running its length on one side, and, secondly, the anus was removed to a place where it was less likely to foul the food currents, namely, at the lower end of the theca near the stem. Bather[100] has suggested that this form also was not completely sessile, but may have moved with the currents, the stem merely keeping the animal above the surface of the substratum; but whether this arrangement is hydrodynamically feasible is another matter. It certainly seems more likely that here again the stem was used as a temporary anchor rather than as a keel.

Another form, contemporaneous with *Dendrocystis*, is *Mitrocystis*. This is rather poorly known, and opinions differ as to its structure. It is interesting for this reason, however: there seem to be two or possibly four side-pores near the junction of theca and stem. There appears to be another opening, most likely a mouth, at the opposite end of the theca, covered by an opercular plate, and there is ornamentation of an adjacent marginal plate which could indicate the site of attachment of a system of tentacles, as in *Gyrocystis*. So if we assume that the mouth opened at the distal end as usual, the pores near the origin of the stem could either be a multiple anus, or, more likely, structures concerned with respiration. The latter function is preferable, in the light of features seen in a rather later heterostele, the Upper Ordovician *Cothurnocystis*. This form is very well known, and has been described in detail by Bather[99]; it is interesting also in connexion with a suggestion about the possible origin of the chordates (p.178). In structure (Fig. 23*d*, *e*, *f*), it has the basic frame of marginal plates supporting an upper and a lower sheet of plates. One of the marginal plates projects forwards on one side, and balancing it is a spine, also projecting forwards, on the other. A large hole at the base of this spine is protected by an operculum. On the under surface there are two strengthening features: first, a strut passes across from one side of the frame to the other; and, secondly, four knobs project downwards from the frame like the studs of a tea-tray, as if to support the theca on the sea-bottom. The feature which makes such an interesting comparison with *Mitrocystis* is the presence of a number (between thirteen and forty-two, depending on size and species) of small elongated grooves (Fig. 23*e*) on the upper surface on one side. Each groove (Fig. 23*f*) is enclosed by a frame consisting of two U-shaped ossicles, the longer of the two having small plates forming a roof between the limbs, the smaller of the two being devoid of cover-plates and surrounding a pore to the interior.

Bather[99] is of the opinion that the larger hole was the anus, the spines adjacent to it being used to direct faecal waste away from the animal, and he thinks that the series of grooves was a ciliated food-intake system, probably assisted in their task by a bellows-like pulsation of the flexible upper side of the theca which could also help to take in and expel water through the anus for respiration. Bather sees this activity as producing a tendency for the theca to pirouette, a turning moment which was possibly offset in some manner by the enlarged marginal spine projecting forwards on the other side. Opposed to this ingenious but rather fanciful theory is Jaekel[102], who prefers to regard the grooves as multiple gonopores and the other vent as combined mouth and anus. While accepting Jaekel's function for the single large hole set apart from the others, one cannot really think of a reason for requiring complex plating over the gonopores, even if so many pores were needed. A third idea comes from Gislén[101], who considers the multiple pores to be gills for the voidance of water taken in through the mouth-anus. Yet despite all these ideas on the function of the slits, nobody seems to have suggested that they housed respiratory organs which could be retracted and covered by plates when the animal was disturbed. One significant thing in both these heterosteles with extra pores is the apparent absence of an anus. It seems likely that all heterosteles, including *Cothurnocystis* used a protrusible system of tentacles for feeding, so that the real mouth and the anus were both borne in the soft tissues of the introvert, not preserved in the fossils. If this were so, then it would be highly desirable to shift the place of gaseous exchange to another part of the body, as far from the anus as possible, which is just what seems to have happened.

The Ophiocistioidea

A few dozen curious fossils, assigned to some five genera, have been found in rocks of Ordovician to Devonian age which their original describer, Sollas[109], thought to be an order, the Ophiocistia, of the ophiuroids. They possess a test (Fig. 24*e*) up to 10 cm in diameter, completely encasing the body, with the mouth in the centre of the under surface; on this surface the plates are fairly regularly arranged into ambulacral and interambulacral columns, but on the aboral surface they do not appear to be arranged in any order. The one outstanding feature of these fossils is the preservation of the tube-feet, which were present on the under surfaces only. It is very rare for these soft structures to be preserved at all from any echinoderms, yet in this group they have been preserved *in situ*, probably because they were exceedingly large and had imbricating plates embedded in their walls. These tube-feet arise from alternating single pores between the perradial and adradial plates of the ambulacra. In two species, *Eucladia johnstoni*[106] and *Euthemon igerna*[109], there is good evidence that oral tube-feet, smaller than the rest and possibly sensory, occurred round the mouth (compare the situation in, say, modern echinoids, p.113), but this division of labour does not seem to be present in the closely related *Sollasina woodwardi*[109].

In all genera the mouth seems to have been bordered by

a flexible peristomial membrane in which many tiny plates were embedded, as in recent echinoids. The mouth itself had five fairly large plates across it, radially arranged, which in the best preserved specimens look as though they

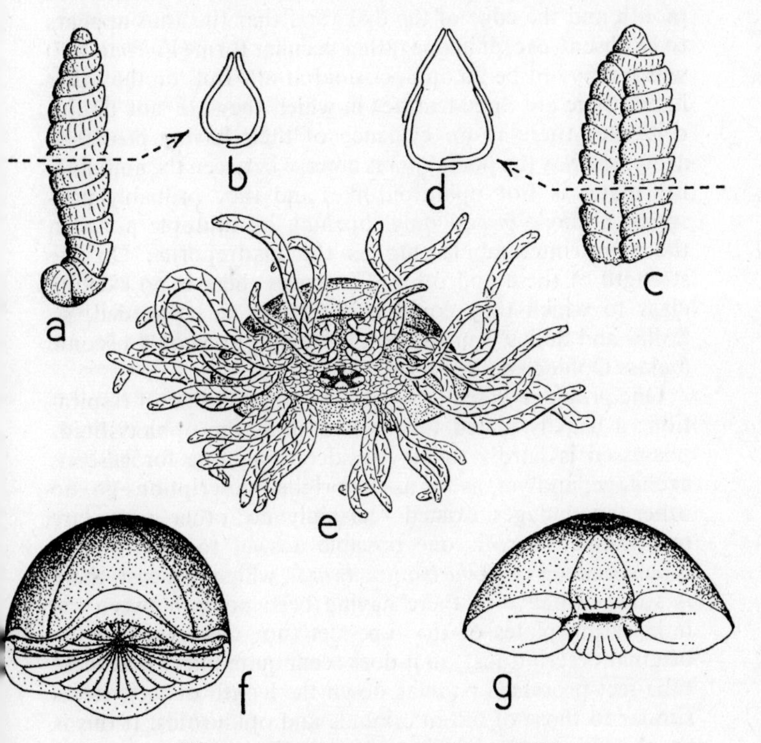

Fig. 24 LESSER-KNOWN FOSSIL GROUPS

a to *d*, MACHAERIDIA.
a Side view, and *b* transverse section of *Lepidocoleus* (U. Ord.), with two columns of plates to the body.
c Side view, and *d* transverse section of *Turrilepas* (M. Sil.), with four columns of plates.
e OPHIOCISTIOIDEA. Reconstruction of *Sollasina* (Sil.–Dev.).
f CYCLOIDEA. Ventro-lateral view of *Cymbionites* (M. Camb.).
g CYAMOIDEA. Ventro-lateral view of *Peridionites* (M. Camb.).

represent a dental apparatus. If this is so, then in this feature too they are more like the echinoids than the ophiuroids.

The features which first suggested ophiuroid affinities are that the madreporite is on the oral surface, between the mouth and the edge of the disk, and that the anus appears to be absent, except in one rather peculiar form (*Volchovia*[107]) which may not be an ophiocistioid at all. But, on the other hand, there are three features in which they are not ophiuroid-like: there is no evidence of their having possessed arms; the way the podial pores emerge between the ambulacral plates is not ophiuroid-like; and they probably possessed a single gonad only, opening by multiple pores in the same interambulacrum as the madreporite. On the strength of these and other differences there is no existing class to which they could conveniently be appended, so Sollas and Sollas[110] proposed that the group should become a class Ophiocistioidea itself.

One problem which arises in this group is that of respiration: a heavily plated tube-foot such as the ophiocistioids possessed is hardly to be considered suitable for gaseous exchange, and yet, as far as the original descriptions go, no other appendages existed—certainly no other pores are mentioned. There is one possible answer to this: on the type specimen of *Sollasina woodwardi*, which is a cast, there is some evidence of there having been pores between the imbricating plates of the tube-feet (not mentioned in the original descriptions), so it does seem quite likely that these tube-feet possessed papillae down the length of their stems similar to those of recent crinoids and ophiuroids. If this is so, then it may explain how they respired.

There is little doubt that the animals were eleutherozoan, and that they moved about on their huge tube-feet, the teeth rasping encrusting organisms from the substratum. The only spines that were present were very tiny ones, of which only the tiny tubercles remain, on the aboral surface. So it seems that this group possesses sufficient features reminiscent of the echinoids (plated theca, teeth, spicule-strengthened peristome) for the two groups possibly to have a common ancestry.

The Machaeridia

This, too, is an enigmatic group, consisting of several hundred Lower Ordovician to Mid-Devonian fossils assigned to four main genera, *Lepidocoleus* (Fig. 24*a*, *b*), *Turrilepas* (Fig .24*c*, *d*), *Plumilites* and *Deltocoleus*, though some others have been provisionally referred to the group. Probably because Bather's *dipleurula* theory of echinoderm ancestry (p.175) was held in high repute since first put forward in the 1920's, and this group was mentioned by him[105] as a possible plated version of this hypothetical ancestor, interest in the group has flared up frequently since Withers's monographic work on them in 1926[113], and much speculation has been voiced about their systematic position. It is only fair to mention that other animal groups, such as the polyplacophoran molluscs, the annelids and the cirripede crustacea, have claimed them, but there is evidence in at least one genus (*Lepidocoleus*) that their plates show the tell-tale single cleavage plane characteristic of echinoderms, and the inner surface of some of the plates shows the familiar reticular pattern. Of course, they could be a separate phylum, sharing these features with the echinoderms, but so far no definitely non-echinoderm animal has been found with this calcite arrangement.

The body of a machaerid (the name means 'little sabre') is shaped like a knife-blade with either a single column of oblong imbricating plates on each side, as in the LEPIDOCOLEIDAE (Fig. 24*a*, *b*), or two columns of subtriangular plates on each side, as in the TURRILEPADIDAE (Fig. 24*c*, *d*), which contains the other three main genera. In some specimens there is a terminal plate at one end of a smaller size and different shape from the others; at the opposite end the body narrows to a point. Scars on the inner faces of the plates of some turrilepadids have been interpreted as sites for the attachment of muscles which served to pull the two sides together, much as a bivalve shell is closed; of course, no other echinoderm shows a body plan anything like this.

Even assuming their affinity with the known echinoderms, much speculation has been enjoyed regarding their systematic position. Bather[104] suggested that they could be parts of the stem of an edrioasteroid (p.149), or a brachiole of a heterostele such as *Dendrocystis*, or they could be parts of a cystoid. 'But,' he writes, 'if it were possible to regard them as such organs isolated from the body, we should be puzzled to find the creature to which they belonged.' A possible answer to this, which, as far as I am aware, has not been made before, is that these structures are isolated tube-feet of their contemporaries, the ophiocistioids. In these, the tube-feet are abnormally large, and some are plated in a manner very similar to the plating of machaerids. Further, there is evidence that the ophiocistioid tube-feet bore papillae from between their plates, and the so-called muscle scars on the inner side of some machaerid plates could conceivably represent the channel in which the coelomic canal feeding these organs ran. There is another feature which supports this idea: the small offset plate at one end could be an axillary plate of a tube-foot which normally lay at an angle, rather than perpendicular, to the test surface.

As far as size-relations are concerned, it is true that most ophiocistioids have tube-feet which are considerably smaller than most machaerids, but there is one known ophiocistioid, *Eucladia johnstoni*, with tube-feet roughly the size of a machaerid's body; unfortunately, in this form the plating arrangement of the tube-feet is unknown. But, of course, there is one outstanding question which is hard to reconcile with this explanation: why have only the tube-feet been preserved and no thecal plates found? Were the plates of these organs held together more firmly than those of the theca? Were the thecae heavily predated for, say, the gonad, while the tube-feet, possibly with distasteful mucus, were rejected? These, and other questons put by the group, cannot be answered without much more evidence; meanwhile, the Machaeridia remain an enigma.

Cycloidea and Cyamoidea

These groups were originally proposed as classes within a new sub-phylum of the echinoderms, the Haplozoa, by their author Whitehouse[112]. Each class was erected on the strength of one fossil, the Cycloidea containing *Cymbionites craticula* (Fig. 24*f*) and the Cyamoidea *Peridionites navicula* (Fig. 24*g*); they both come from the lower part (*Xystridura* zone) of the Middle Cambrian of Queensland, *Cymbionites* ranging through forty feet and appearing in colossal numbers in some bands, while *Peridionites*, twenty-four feet above, ranges through five feet and is far less numerous. They have been included in the phylum solely because their plates are apparently single crystals of calcite, but they are so different from previously known echinoderms that they cannot be included in any known class. *Cymbionites* consists of a dome-shaped, radially symmetrical theca about 10 mm long composed of a radial series of wedge-shaped plates, usually five in number; the concave under surface is ridged and grooved in a manner which suggested to Whitehouse that the soft parts of the body, with straight gut and coelomic pouches, etc., were all contained beneath the theca, held in place by radially running ligaments inserted into grooves visible on the under surface of the fossils. *Peridionites* also has a dome-shaped theca of five plates, but they are not arranged pentamerally and in consequence the symmetry is biradial. The under surface is again concave, but here Whitehouse sees the possibility of it having housed a bilateral animal, possibly close to the hypothetical *dipleurula* ancestor of the phylum (p.175).

It need hardly be said that on such slim evidence as the structure of their very peculiar tests *Cymbionites* and *Peridionites* have been by no means universally accepted as representative of a new group of the echinoderms, though it is rather alarming to see how many textbooks, particularly of palaeontology, have seized on the group as giving fossil evidence for the *dipleurula* theory. It has been suggested that they belong to the Eocrinoidea (p. 29), or alternatively that they are not echinoderms at all, though if not they join the Machaeridia in being possibly a non-echinoderm group showing what has hitherto been regarded as a unique echinoderm feature, namely, plates composed of a single calcite crystal.

13

THE PHYLOGENY OF THE ECHINODERMATA

THE unfortunate thing about building a phyletic tree of any animal phylum is that one can never trust the early fossil record: though it is perfectly true that only a sequence of fossils can indicate the true course of evolution, yet the vagaries of fossilization may have caused primitive groups to appear in the record later than less primitive ones. If what we surmise about the invertebrate relations of the echinoderms (see Chapters 1 and 14) is anywhere near the truth, then the groups which can claim to be closest to the ancestral echinoderms are either the cystoids or the heterosteles, on the grounds that some of their early members at least probably had soft-tentacled introverts for feeding, and did not exhibit pentamerous symmetry But the trouble is that these forms appear long after the first crinoids, and even later than the first edrioasteroids, so that on palaeontological grounds we must assume either that these non-pentamerous groups are secondarily modified or that they and the groups derived from them have reached an echinoderm grade independently of the other classes and that the stages by which they did so are missing from the record.

So we must say that though the EOCRINOIDEA are the first echinoderms to appear, and doubtless gave rise to the PARACRINOIDEA and true CRINOIDEA, their exact relationships to the somewhat later CYSTOIDEA and HETEROSTELEA are not clear.

Of the subsequent groups, it is fairly clear that the

BLASTOIDEA were an offshoot from the cystoids (Fig. 25), a line which went no further. Though in thecal structure it has been said[1] that they resemble the crinoids, they possess other features, such as the hydrospire system and the ambulacral plate arrangement, which makes a connexion between these two difficult to envisage. Further, the path from the pentamerous cystoids to the blastoids is fairly well sign-posted by the transition-form *Blastoidocrinus* (Fig. 21*b*).

The crinoids went on to attain fantastic success in the Palaeozoic. It seems possible that early in their history they gave rise to the EDRIOASTEROIDEA. It should be recalled that in the crinoids a spicule-strengthened tegmen covered the oral surface of the theca, with the mouth at or near the centre. If one imagines a primitive crinoid, such as *Stephanocrinus*, stripped of its arms and pinnules, so that only those parts of the ambulacral grooves on the tegmen are left, then one has a structure not far removed from the basic pattern of such edrioasteroids as *Stromatocystis* (Fig. 22*c*).

So far, all echinoderms we have mentioned have fed entirely on the rain of detrital matter falling to the sea-floor from the waters above; they have been almost entirely rooted to one spot, their success depending on their ability to command a catchment area sufficient for their needs. But now, while the crinoids continue to exploit this method, a dramatic change occurs in the edrioasteroid line, and maybe others too, which involves turning the animal completely upside down, so that the prehensile, and probably by now muscular, tube-feet can be used for locomotion. No longer is the animal rooted, plant-like, to one spot, but is at last eleutherozoic, free to wander over the substratum, with the consequent ability to exploit fresh ecological niches. This means, of course, that the problem of feeding must be solved anew, and one can say that each of the four living eleutherozoan classes have solved the problem in a different way.

First came the ASTEROIDEA, at some time in the Ordovician, but whether they came from the edrioasteroids or direct from the crinoids is debatable. It would take little

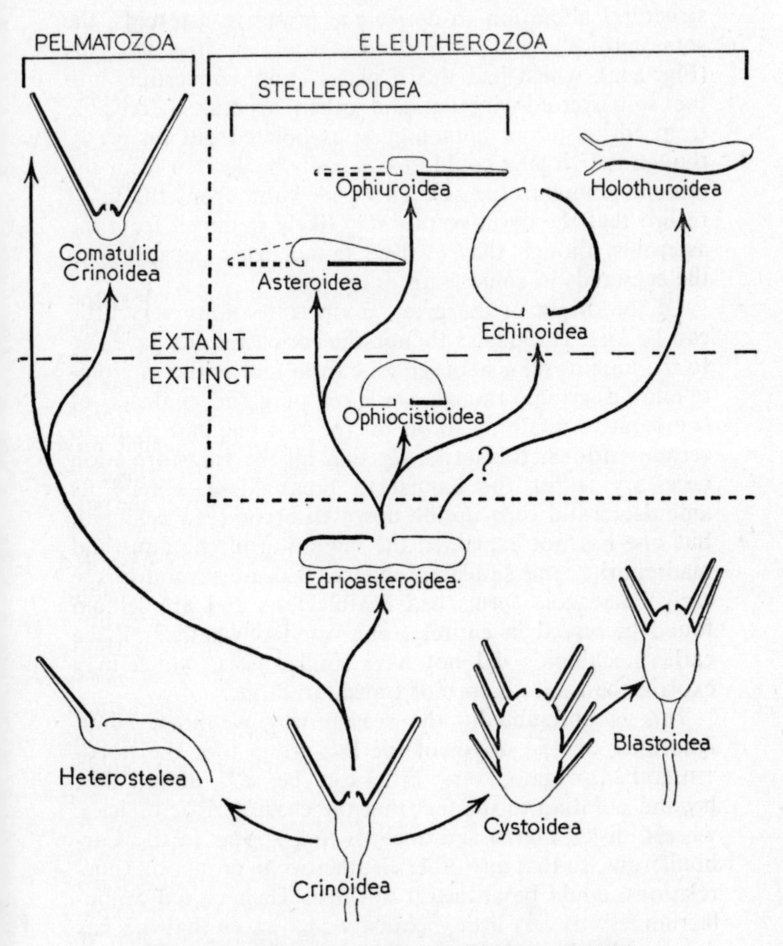

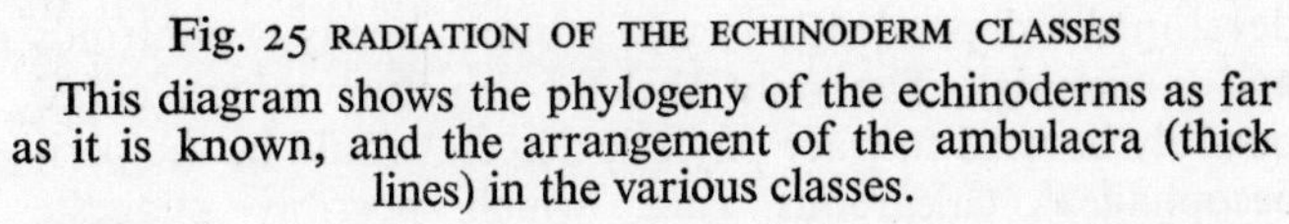
Fig. 25 RADIATION OF THE ECHINODERM CLASSES

This diagram shows the phylogeny of the echinoderms as far as it is known, and the arrangement of the ambulacra (thick lines) in the various classes.

structural alteration to derive the primitive asteroids, the somasteroids, from such edrioasteroids as *Stromatocystis* (Fig. 22*c*), which had thecal plates easily convertible into the somasteroid virgalia, and other structures derivable from edrioasteroid homologues, as pointed out on p. 154, though the virgalia could just as easily be derived from the crinoid pinnulars (p. 24). Then, we know from the fossil record that the OPHIUROIDEA very likely stemmed from the asteroids, though they evolved tantalizingly parallel with the echinoids in some features (p. 59).

Of the origin of the great group ECHINOIDEA very little can be said. It looks as though the edrioasteroids gave rise to the urchins on a stem separate from that of the asteroid-ophiuroid group, though we have only the evidence of comparative anatomy to go on (p. 71); one can point to certain edrioasteroids and say that all the transformation necessary is for the animal to break loose, extend its ambulacra and turn upside down to become an echinoid, but one has not explained the migration of the anus and madreporite, and similar problems. Because several of the early Palaeozoic forms had flexible tests and are seldom found preserved in entirety, one wonders whether all the earliest echinoids did not have similar tests, which may explain the total absence of transition forms.

This is undoubtedly the reason why we must admit ignorance on the origin of the last group too, the HOLOTHUROIDEA. Again, very little can be said about this, beyond pointing to the fact that the characteristic ossicles, wheels, disks, hooks and anchors first appear in the Carboniferous, so that any other eleutherozoan group, on time-relations, could be ancestral to them. Their closed ambulacrum tells us very little, because we have seen that this has developed independently in two groups already (ophiuroids and echinoids). The only skeletal structure which could possibly be used to trace phyletic descent would be the peri-oesophageal calcareous ring, which possibly suggests affinity with either asteroids or echinoids, rather than crinoids.

14

RELATIONS OF THE ECHINODERMS TO OTHER PHYLA

Invertebrate relations and origin of the Echinodermata

To put the echinoderms into perspective in the animal kingdom we need to trace the history of the invertebrates, as far as can be done with our limited knowledge, from quite early times, at the grade when the annelid worms arose from their forbears. The radiation into the main phyla, however, occurred in the Precambrian, long before we can expect to find much fossil evidence. Clearly, there are various ways in which data derived from the study of such evidence as is available to us from present-day representatives can be arranged so as to give a plausible picture of evolutionary relationship. What has to be decided is which arrangement is the most satisfactory in using the greatest number of relevant facts, at the same time leaving the fewest unaccounted for. One of the most thought-provoking problems in this kind of phyletic speculation is the relative weighting one can give to the various types of evidence. Is one to consider developmental similarity as necessarily more valuable in tracing descent than anatomical? Which anatomical features in an animal are least likely to be influenced by its specific environment and which can be expected to be reasonably static for our purpose? And so on. What I have attempted to do in the pages that follow is to present what I consider to be the most plausible arrangement of the phyla adjacent to the echinoderms in the phyletic tree, together with the most important evidence. Of necessity, much is left out, but anybody wishing to trace this fascinating subject further can do so through the references quoted.

Now, to return to the annelid grade of invertebrate organization: this was the stage at which metameric segmentation was seized upon for the first time, probably because of the increased locomotor efficiency it bestowed. It was also a stage at which a peculiar form of development appeared, a type of cleavage which though basically similar to the spiral pattern possessed by some of the earlier invertebrates, the platyhelminths and aschelminths, came to show a recognizably unique cross-like pattern of cells on the dorsal surface of the blastula which was different from anything previously seen. This pattern is known as the *annelid cross*. Subsequent development in the annelids leads to a larva called the *trochophore*, which exhibits a fairly constant pattern among all the annelids, that is, with much less variation of form than is found in the larvae of the echinoderms. The larva is roughly spherical, with an equatorial ring of cilia and an apical sense organ, a mouth on one side just below the ciliated ring, and an anus near the base.

Many textbooks still place the Sipunculida as a class of the annelids, principally because they too show spiral cleavage, have the peculiar annelid-type blastula with its cross, and disperse by means of a trochophore larva not unlike that of the annelids. But, on the other hand, there is no trace of segmentation in the adult or in the larva, and they have a recurved gut, a feature possessed by no annelid, even those burrowing forms in which one would expect such an arrangement to be an advantage. Other aspects of their adult anatomy also belie such a close connexion, particularly the possession of a peculiar tentacle arrangement round the mouth. This varies enormously in different sipunculids, but its basic anatomy appears to hold a clue to the subsequent evolution of invertebrate phyla and particularly the origin of the echinoderm tubular coelomic system. In general, it consists of a number of tentaculated outgrowths of the body wall, heavily ciliated externally, which provide a catchment funnel for food. The cavities of the tentacles lead into a circum-oesophageal vessel close to the mouth, to which one or more vesicles, the so-called 'compensation sacs', are connected, hanging in the body cavity alongside the gut. The walls of these sacs are muscular, and by their contraction, as one would expect, the tentacles round the mouth are extended. It should be quite evident from this brief account of Sipunculida that in their basic adult anatomy they closely resemble the picture of a primitive adult echinoderm outlined in Chapter 1. But this is by no means the whole story: one expects a good deal of convergence when two groups adopt similar modes of life, and there are features which do not quite fit the picture.

The blastopore of the sipunculid larva is the site of, or close to, the future adult mouth, that is, the phylum is *protostomatous*; but in the echinoderms (p.120) the blastopore becomes the adult anus (*deuterostomatous*), and these differences in mode of development seem to represent divergence at an early stage in evolutionary history. A protostomatous condition is retained in a line of minor coelomate phyla sometimes lumped together as the *Lophophorata*. These phyla are the Phoronida, Ectoprocta and Brachiopoda. All are characterized by the presence in the adult of a hollow-tentacled feeding funnel round the mouth, the *lophophore*, the cavities of whose tentacles are coelomic and connected by a ring or loop round the oesophagus—an arrangement very similar, in fact, to that in sipunculids. Furthermore, the lophophorates have trochophore larvae, metanephridia and, in general, a schizocoelous coelom (body cavity formed by splitting of the mesoderm, though this is secondarily variable), all of which they share with the sipunculids. So one can fairly safely suggest a connexion between the Sipunculida and the Lophophorata.

There is some evidence that the lophophorate body is subdivided into three regions: protosome, mesosome and metasome, though the protosome is generally very much reduced. The lophophore itself is borne by the mesosome, and there is a strong partition between this division of the body and the metasome. Such division of the body is also shown by phyla on what is almost certainly a second line derived from the sipunculids, including the Chaetognatha, Pogonophora and Hemichordata, and, of course, the Echinodermata (p. 120). Further, in the latter two groups there are tentacled outgrowths from the middle body division, as in the lophophorates; in the echinoderms the ambulacral system of tubes represents this.

Of these four phyla, certain features place the Chaetognatha, the arrow-worms found in the plankton, rather apart from the others. Chief among these features are the mode of formation of the coelom, which is by a rather modified form of enterocoely, and the fact that only during their development do they profess their coelomate and tripartite arrangement at all; subsequently the coelom is filled in with mesenchyme and a secondary cavity is formed, closely resembling that of the pseudocoelomate phyla such as the Nematoda. It has been suggested[119] that the Chaetognatha lie at the base of the line leading towards the echinoderms and chordates.

The remaining three phyla, Pogonophora, Hemichordata and Echinodermata, are united by possession of a number of features, among which we may mention the formation of the coelom by enterocoely, the retention of the blastopore, when present, as the site of the future anus (deuterostomy), and, in those with indirect development, a dipleurula-type larva at some stage. Also in all three there is a tendency to reduce the right anterior coelomic pouch during development: at least, one can say that this is so in the hemichordates and echinoderms, but for the pogonophores one can go no further at present than to say that the protocoel is probably single. In all three there is a heart vesicle (pericardial sac in some pogonophores) which may represent the right anterior coelom. The adult pogonophores, with their complete absence of alimentary system, are to be regarded as highly specialized offshoots of the line leading to the echinoderms. Very little is known about the pogonophore larva, beyond the fact that it is bilateral, has no blastopore and shows some signs of an incipient gut which never develops very far. The larvae of the other two have been well studied, however, and show fairly conclusively that the hemichordates and echinoderms are closely linked[118]. The hemichordate larva, called a *tornaria*, has the same basic structure as the dipleurula of an echinoderm, with the difference that there is an extra ciliated band posteriorly, the telotroch. In other details there is remarkable similarity, such as the presence of a hydroporic canal associated with the left side in both. This similarity is confirmed in adult features, too, and Grobben[118] has shown the principal transformations necessary to convert a pterobranch hemichordate into an echinoderm, a derivation which requires chiefly the reduction of the right middle coelom (hydrocoel) and the growth round the oesophagus of the remaining hydrocoel pouch; the hemichordate cephalic shield he considers as the stalk and attachment disk of pelmatozoa and asteroids respectively (Fig. 26).

There have been several attempts to construct a hypothetical common ancestor of all the echinoderms and the hemichordates, the best known of which is that of Semon[123], developed later by Bather[1]. Bather sees the ancestor as resembling the *dipleurula* larval stage, but with a full complement of coelomic pouches on either side of the gut, as in Fig. 26*a*; it was a creeping animal, with anterior sense organ and ventral mouth and anus. Each axohydrocoel (p. 120) opened by a hydropore dorsally. A postulate of Bather's theory is that the ancestor at some stage became attached by the front end (as do modern larval pelmatozoa and asteroids) and underwent torsion in such a way that the left axohydrocoel was reduced and the right one came to surround the gut. Bather shows the way this might have happened in the pelmatozoa, with mouth uppermost (Fig. 26*c*), and in asteroids, with mouth to one side (Fig. 26*d*). Although the movement of the internal organs of the pelmatozoa on this theory looks drastic, a very important piece of supporting evidence for the idea is found in the fossil cystoid *Aristocystites* (p.138 and Fig. 19*b*) which has a structure closely resembling that of Bather's hypothetical primitive pelmatozoan after torsion. As was mentioned in Chapter 11, it was probably at about this stage in the evolution of the group that the advantages of a rigid skeleton of plates was seized upon, followed by the adoption of pentamery.

Bather is at a loss to explain the disappearance of the right axohydrocoel. In fact, nobody has given a satisfactory explanation of the mystery. Certainly, the symmetry of the dipleurula if it existed at all was lost very early, since no fossil with two hydropores has ever been found. Gemmill[27] has reviewed the occurrence of anomalous asteroid larvae with double hydropores, which occur in up to thirty-three per cent of individuals in some larval cultures, and has shown that later in life the right hydropore is resorbed; this curious state of affairs has been used to suggest that the situation did once exist, that the odd specimens found are a sort of throwback.

The subsequent development of the eleutherozoan type, with the mouth first formed on one side of the larva (Fig. 26*d*), was probably by a less drastic torsion, with the consequent lack of twisting of the gut, retained in the asteroids and ophiuroids. The resulting picture is not unlike that of another hypothetical ancestor of the echinoderms (but not shared with the hemichordates), the *pentactula*, first suggested again by Semon[123], but developed later by Bury[116]. In this theory the original echinoderm already has five tentacles, the lumina of which stem from a circum-oral coelomic vessel. To me, it seems simplest to regard the pentactula as a possible later stage of the ancestral echinoderm, the larva of which was the dipleurula, and as representing an ancestor already well within the confines of our definition of an echinoderm. But there seems to be sufficient evidence to show that there is no need to regard the original echinoderm (p.20) as pentamerous. One should describe it as a sessile animal with the mouth directed upwards and any number of food-gathering tentacles, probably with a supporting skeleton, arranged in a ring round it. The lumina of these tentacles join a circular vessel round the first part of the gut, and this vessel has an opening to the

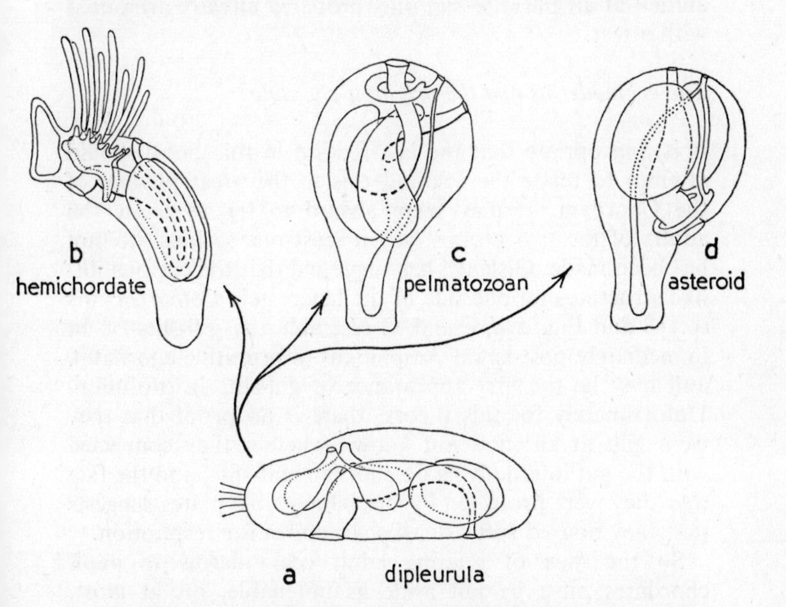

Fig. 26 POSSIBLE DERIVATION OF PTEROBRANCH HEMICHORDATES, PELMATOZOAN AND ELEUTHEROZOAN ECHINODERMS FROM THE ANCESTRAL DIPLEURULA

(modified after Bather and Grobben)

a The *dipleurula*, drawn upside down when compared with Fig. 16*a*, showing the paired axohydrocoels, each with its own hydropore, and paired somatocoels.

b Pterobranch, showing migration of anus, single anterior coelom and paired middle coelom, with branches to the tentacles.

c Pelmatozoan, showing changes envisaged by Bather in bringing the mouth to the top. One axohydrocoel is lost; the axocoel part of the other forms the parietal canal, lost later in ontogeny.

d Eleutherozoan, in which less drastic torsion has occurred.

exterior near the mouth. The anus opens to one side of the theca. The dipleurula-concept may represent a possible *larval stage* common to the echinoderms and hemichordates, but the original echinoderm was probably not a free-living animal at all but a sessile one, probably already armoured with a test.

The echinoderms and the origin of chordates

It is appropriate that the last section in this book should attempt to relate the echinoderms to the great phylum of the Chordata. It may seem absurd to try to relate the adults of the two groups, yet at least one serious attempt has been made. Gislén[131] has suggested that the curious slit-like structures on one side of the heterostele *Cothurnocystis* (p.158 and Fig. 23*d*, *e* and *f*) resemble the gill-slits of an immediately post-larval Amphioxus (a primitive chordate), and may be the first appearance of gill-slits in evolution. Unfortunately for this theory, there is no proof that they were gills at all: it is not known whether they connected with the gut internally as do chordate gill-slits, and the fact that they were protected by special movable plates suggests that they housed soft extensile structures for respiration.

So the idea of relating adult echinoderms to adult chordates must be put aside as untenable, or, at most, unlikely. How, then, did the fish-like organization of the primitive adult chordate arise from the invertebrate stock? How did the unique features of dorsal nerve cord, ventral heart, pharyngeal gill-slits and notochord, followed by vertebral column, jaws and so on, come into being? The clue to these questions probably lies in an interesting embryological phenomenon known as *neoteny*, in which, by the acceleration of development of the gonads, an animal becomes sexually mature while still retaining the larval body form. In other words, the animal never undergoes metamorphosis. Although this process can be induced phenotypically (that is, by an influence of the environment) it also seems highly probable that it has been utilized during evolution to bring about various new groups, some of which have shown tremendous potentiality for exploiting new habitats thus opened to them. It is unlikely that a new group could *suddenly* be produced in this way by one drastic genetic change, but more probable that the onset of sexual maturity is *gradually* brought forward during development.

We have seen in Chapter 10 that the early dipleurula larva of echinoderms has a ciliated band running along its sides near the dorsal surface; the bands from the two sides join just in front of the anus and also near the mouth, where a continuation of each side, the *adoral* portion, passes a little way into the larval oesophagus on its ventral side. The point to be borne in mind here is that every ciliated band, if all its cilia are to work in concert for locomotion, must be underlain by some sort of nerve tract. So we have, in the auricularia, a strong aggregation of nerve tissue near the dorsal surface on each side. The theory of chordate origin first propounded by Garstang[129] in 1894 proposes that these two bands moved further towards the mid-line and rolled in on themselves to form a dorsal nerve tube. The reason for such a movement may have been that strong lateral blocks of swimming muscles were required for locomotion in place of the rather weak method of ciliary beat, particularly if selection was acting to increase the time the larva was in the plankton and probably also its final size. Such an increase in size would immediately raise two problems: first, there would be a need for extra support, particularly to maintain a constant length, and this may well have been solved by the incorporation of a rod, the *notochord*, down the back of the animal. We can be fairly certain that this rod consisted originally, as it does now, of heavily vacuolated cells whose turgidity provided the necessary strength, rather as a plant is supported by the pressure of fluid in its cells. Secondly, with increase in size and activity, gaseous exchange by diffusion over the whole body surface could no longer cope with the respiratory needs of the animal, so gill-slits were provided in the first part of the gut, the *pharynx*, so that water could be brought into the mouth, filtered to collect food particles and passed across a surface in close contact with some sort of vascular fluid. This current was probably helped into the gut by the retained part of the ciliated band inside the mouth, now called the *endostyle*. A larva of this type would be well suited to exploit the rich harvest of plankton near the sea-surface, so it would be to its advantage to delay metamorphosis into a sessile adult for as long as possible—for ever, if it could. And this is where the process of neoteny comes in: selection would favour the gradual advance in the onset of sexual maturity, so that finally an adult animal existed, the *protochordate*, which had a dorsal neural tube, gill-slits, a notochord and an endostyle. Further increase in size of this form would mean that the notochord became inadequate to counter the stresses of muscular movement, and a bony structure, the vertebral column, was incorporated around it.

So we have now reached a stage recognizable as fish-like. There is no general agreement as to which living forms among the primitive chordates are closest to the ancestral pattern, but the views of Berrill[124], who regards the tadpole of an ascidian (sea-squirt) as occupying this position, appear to find widest acceptance. But whatever are the finer points concerning the course actually taken during the early evolution of the uniquely successful chordates, it is clear that the early echinoderms represented the spring-board from an invertebrate condition to the vast potentialities of the vertebrates.

APPENDIX

A CLASSIFICATION OF THE ECHINODERMATA

This list includes all the genera mentioned in this book.

† extinct groups or genera

(Br.) those occurring in British waters

SUB-PHYLUM PELMATOZOA

CLASS CRINOIDEA

SUB-CLASS EOCRINOIDEA† *Eocystites*, *Macrocystella*

SUB-CLASS PARACRINOIDEA† *Comarocystites*

SUB CLASS EUCRINOIDEA

Order Inadunata† *Dendrocrinus*, *Hybocystis*, *Petalocrinus*

Order Flexibilia† *Protaxocrinus*, *Ichthyocrinus*

Order Camerata† *Platycrinites*, *Barrandeocrinus*

Order Articulata *Pentacrinites*†, *Hyocrinus*, *Antedon* (Br.), *Ptilocrinus*, *Neocomatella* (Br.)

CLASS CYSTOIDEA†

Order Diploporita *Aristocystites*, *Holocystites*, *Eucystis*, *Glyptosphaerites*, *Proteocystis*, *Fungocystis*, *Protocrinus*, *Proteroblastus*

Order Rhombifera *Echinosphaerites*, *Lepadocrinus*, *Heliocrinus*, *Callocystites*, *Cystoblastus*

CLASS BLASTOIDEA†

Order Parablastoidea *Blastoidocrinus*

Order Coronata *Stephanocrinus*

Order Eublastoidea *Codaster*, *Pentremites*, *Eleutherocrinus*, *Zygocrinus*, *Pterotoblastus*, *Orbitremites*, *Orophocrinus*

CLASS HETEROSTELEA *Dendrocystis*, *Gyrocystis*, *Trochocystis*, *Mitrocystis*, *Cothurnocystis*

CLASS EDRIOASTEROIDEA†

Family Agelacrinidae *Agelacrinus*, *Pyrgocystis*, *Stromatocystis*, *Hemicystis*, *Lepidodiscus*
Family Cyathocystidae *Cyathocystis*
Family Edrioasteridae *Edrioaster*, *Dinocystis*
Family Steganoblastidae *Steganoblastus*

SUB-PHYLUM ELEUTHEROZOA

CLASS ASTEROIDEA

SUB-CLASS SOMASTEROIDEA

Order Platyasterida *Villebrunaster*†, *Chinianaster*†, *Archegonaster*†, *Platasterias*
Order Hemizonida† *Urasterella*, *Cnemidactis*, *Arthraster*

SUB-CLASS EUASTEROIDEA

Order Phanerozonia *Luidia* (Br.), *Astropecten* (Br.), *Porania* (Br.), *Pectinaster*, *Petraster*†, *Hudsonaster*†, *Xenaster*†
Order Spinulosa *Solaster*, *Henricia* (Br.), *Asterina* (Br.), *Anseropoda* (Br.), *Pteraster*
Order Forcipulata *Asterias* (Br.), *Stichastrella* (Br.), *Heliaster*, *Brisinga*

CLASS OPHIUROIDEA

Order Stenurida† *Pradesura*, *Eophiura*, *Palaeura*
Order Auluroidea† *Aspidosoma*
Order Ophiurida† *Stenaster*, *Taeniaster*, *Lapworthura*
Order Ophiurae *Ophiura* (Br.), *Ophiocomina* (Br.), *Ophiothrix* (Br.), *Ophiopsila* (Br.), *Ophiactis* (Br.), *Amphiura* (Br.), *Acrocnida* (Br.), *Amphipholis* (Br.), *Ophiotholia*
Order Euryalae *Asteronyx*, *Astroschema*, *Gorgonocephalus*

CLASS OPHIOCISTIOIDEA† *Sollasina*, *Euthemon*, *Eucladia*, *Volchovia*

CLASS ECHINOIDEA

SUB-CLASS PERISCHOECHINOIDEA

Order Bothriocidaroida† *Bothriocidaris*
Order Echinocystitoida† *Aulechinus*

Order Palaechinoida† *Palaechinus*, *Melonechinus*, *Lepidesthes*
Order Cidaroida *Cidaris*, *Miocidaris*†

SUB-CLASS EUECHINOIDEA
Super-order Diadematacea *Pygaster*†, *Echinothuria*, *Diadema*
Super-order Echinacea *Echinus* (Br.), *Paracentrotus* (Br.), *Colobocentrotus*, *Toxopneustes*
Super-order Gnathostomata
Order Holectypoida *Holectypus*†, *Galeropygus*†, *Echinoneus*, *Micropetalon*
Order Clypeasteroida *Rotula*, *Echinocyamus* (Br.), *Clypeaster*
Super-order Atelostomata
Order Holasteroida *Holaster*†
Order Nucleolitoida *Nucleolites*†
Order Cassiduloida *Cassidulus*†
Order Spatangoida *Spatangus* (Br.), *Echinocardium* (Br.), *Brissopsis* (Br.), *Schizaster*, *Pourtalesia*, *Echinosigra*, *Hagenowia*†

CLASS HOLOTHUROIDEA
Order Dendrochirota *Cucumaria* (Br.), *Pseudocucumis* (Br.), *Thyone* (Br.), *Psolus*, *Rhopalodina*, *Scotoplanes*
Order Aspidochirota *Holothuria* (Br.)
Order Elasipoda *Pelagothuria*, *Peniagone*, *Psychropotes*
Order Molpadonia *Molpadia*, *Caudina*
Order Apoda *Synapta*, *Leptosynapta* (Br.), *Labidoplax* (Br.)

INSERTAE SEDIS

CLASS MACHAERIDIA†
Family Lepidocoleidae *Lepidocoleus*
Family Turrilepadidae *Turrilepas*, *Plumilites*, *Deltocoleus*

SUB-PHYLUM HAPLOZOA†

CLASS CYCLOIDEA *Cymbionites*

CLASS CYAMOIDEA *Peridionites*

BIBLIOGRAPHICAL REFERENCES

After a section listing general works covering the whole phylum, the references are divided into sections corresponding approximately to the chapters. Though all the references are numbered consecutively, each section is arranged in alphabetical order of authors. This list has been kept to a minimum and includes only those which the author thinks will be of interest to a student requiring further information on those topics discussed in this book; it is not intended to be a list of sources from which material has been taken.

* Denotes a review paper, or a work with comprehensive bibliography.

GENERAL WORKS ON ECHINODERMS

1 *Bather, F. A., Gregory, W. K., and Goodrich, E. S., 1900. 'The Echinoderma' in *A treatise on zoology*, ed. E. Ray Lankester. London.

2 *Boolootian, R. A., and Giese, A. C., 1958. *Biol. Bull. Wood's Hole*, *115*, 53–63. (Coelomic corpuscles of echinoderms)

3 Cuénot, L., 1891. *Arch. Biol. Paris*, *11*, 313–680. (Morphology and histology of echinoderms)

4 *Cuénot, L., 1948. 'Echinodermes' in *Traité de zoologie*, ed. P-P. Grassé. Paris.

4*a* Fell, H. B., in prepn. *The Echinoderms*. London.

5 Forbes, E., 1841. *A history of British . . . Echinodermata*. London.

6 Gislén, T., 1924. *Zool. Bidr. Uppsala*, *9*, 1–316. (Morphology, behaviour and mode of life of echinoderms)

7 *Hyman, L. H., 1955. *The invertebrates*, vol. IV. *Echinodermata*. New York.

8 MacBride, E. W., 1906. 'Echinodermata' in *The Cambridge natural history*, ed. Harmer and Shipley. London.

9 Moore, R. C., Lalicker, C. G., and Fischer, A. G., 1952. *Invertebrate fossils*. New York.

10 *Regnéll, G., 1960. *Palaeontology*, 2, 161–79. (Lower Palaeozoic echinoderm faunas of Britain and Scandinavia)

11 Schrock, R. R., and Twenhofel, W. H., 1953. *Principles of invertebrate paleontology*. New York.

THE CRINOIDS

12 Carpenter, P. H., 1884. '*Challenger*' *Rep., Zool., 26*. (The stalked crinoids)

13 Chadwick, H. C., 1907. *L.M.B.C. Memoir, 15*. (*Antedon*)

14 *Clark, A. H., 1915–50. *Bull. U.S. nat. Mus., 82* (Monograph of the existing crinoids)

15 Dimelow, E. J., 1958*a*. *Nature, Lond., 182*, 812. (Pigments in *Antedon*)

16 Dimelow, E. J., 1958*b*. *Thesis for Ph.D., Univ. of Reading*. (Biology of *Antedon* and *Neocomatella*)

17 Moore, A. R., 1924. *J. gen. Physiol., 6*, 281–8 (Nervous coordination in *Antedon*)

18 Moore, R. C., 1950. *Int. geol. Congr. Rep. 18th Sess., Gt. Brit., 1948, 12*, 27–53. (Evolution of crinoids)

19 *Moore, R. C., and Laudon, L., 1943. *Geol. Soc. Amer., Special papers 46*. (Classification and evolution of palaeozoic crinoids)

20 *Springer, F., 1920. *Smithson. Inst. Pub. 2501*. (Monograph of the Flexibilia)

THE ASTEROIDS AND OPHIUROIDS

21 Anderson, J. M., 1953. *Biol. Bull. Wood's Hole, 105*, 47–61. (Histology of pyloric stomach in *Asterias*)

22 Anderson, J. M., 1954. *Biol. Bull. Wood's Hole, 107*, 157–73. (Histology of cardiac stomach in *Asterias*)

23 Binyon, J., 1961. *J. Mar. biol. Ass. U.K., 41*, 161–74. (Salinity tolerance of *Asterias*)

24 Binyon, J., 1962. *J. Mar. biol. Ass. U.K., 42*, 49–64. (Ionic regulation in *Asterias*)

25 Chadwick, H. C., 1923. *L.M.B.C. Memoir, 25*. (*Asterias*)

26 Fell, H. B., 1960. *Synoptic key to the genera of ophiuroids.* Victoria Univ., Wellington, N.Z.

27 Gemmill, J. F., 1914. *Phil. Trans.*, B, *205*, 213–94. (Development and adult structure of *Asterias*)

28 Gordon, I., 1929. *Phil. Trans.*, B, *217*, 289–334. (Skeletal development in *Leptasterias*)

29 Lavioe, M. F., 1956. *Biol. Bull. Wood's Hole*, *111*, 114–22. (How starfishes open bivalves)

30 Smith, J. E., 1937. *Phil. Trans.*, B, *227*, 111–73. (Nervous system of *Marthasterias*)

31 Smith, J. E., 1950. *Phil. Trans.*, B, *234*, 521–58. (Nervous system of *Astropecten*)

32 Smith, J. E., 1950. *Symp. Soc. exp. Biol.*, *4*, 196–220 (Nervous mechanisms underlying behaviour in starfishes)

33 *Spencer, W. K., 1914–40. *Palaeontogr. Soc.*, *67*, *69*, *70*, *71*, *74*, *76*, *79*, *82*, *87*, *94*. (British palaeozoic asteroids and ophiuroids)

34 Spencer, W. K., 1950. *Geol. Mag.*, *87*, 393–408. (Faunal succession of palaeozoic asteroids)

35 Spencer, W. K., 1951. *Phil. Trans.*, B, *235*, 87–129. (Somasteroids and early ophiuroids)

THE ECHINOIDS

36 Chadwick, H. C., 1900. *L.M.B.C. Memoir*, *3*. (*Echinus*)

37 *Durham, J. W., and Melville, R. V., 1957. *J. Pal.*, *31*, 242–72. (Classification of the echinoids)

38 Forster, G. R., 1959. *J. Mar. biol. Ass. U.K.*, *38*, 361–7. (Quantitative ecology of *Echinus*)

39 Hawkins, H. L., 1919. *Phil. Trans.*, B, *209*, 377–480. (Echinoid ambulacra)

40 Hawkins, H. L., 1927. *Biol. Rev.*, *3*, 281–3. (Some problems in the evolution of echinoids)

41 Hawkins, H. L., 1943. *Quart. J. geol. Soc. Lond.*, *99*, lii–lxxv. (Evolution and habit of echinoids)

42 Lovén, S., 1833. *K. svenska VetenskAkad. Handl.*, *19*, 1–95. (Anatomy of *Pourtalesia* and other echinoids)

43 Moore, H. B., 1936. *J. Mar. biol. Ass. U.K.*, *20*, 655–71 (Biology of *Echinocardium*)

44 *Mortensen, T., 1928–51. *A monograph of the Echinoidea.* 5 vols. Copenhagen.

45 Nichols, D., 1959*a*. *Phil. Trans.*, B, *242*, 347–437. (Morphology and mode of life of recent and fossil spatangoids)

46 Nichols, D., 1959*b*. *Syst. Assoc. Publ.*, *3*, 61–80. (Mode of life of recent and fossil spatangoids)

47 *Otter, G. W., 1932. *Biol. Rev.*, *7*, 89–107. (Rock-boring echinoids)

THE HOLOTHUROIDS

48 Chun, C., 1900. *Aus den Tiefen des Weltmeeres.* Berlin. (Anatomy of *Pelagothuria*)

49 Clark, H. L., 1907. *Smithson. Contr. Knowl.*, *35*, 6–231. (Apoda and Molpadonia)

50 Croneis, C., and McKormack, J., 1932. *J. Pal.*, *6*, 111–48. (Fossil holothuroids)

51 Deichmann, E., 1930. *Bull. Mus. comp. Zool., Harvard*, *71*. (Holothuroids of W. Atlantic)

52 Endean, R., 1957. *Quart. J. micr. Sci.*, *98*, 455–72. (Cuvierian organs of *Holothuria*)

53 Endean, R., 1958. *Quart. J. micr. Sci.*, *99*, 47–60. (Coelomocytes of *Holothuria*)

54 Hamann, O., 1883. *Z. wiss. Zool.*, *39*, 145–90. (Histology of some holothuroids)

55 Théel, H., 1882. '*Challenger*' *Rep., Zool.*, *4*. (Elasipoda)

PENTAMERY AND THE ECHINODERM SKELETON

56 Breder, C. H., 1955. *Bull. Amer. Mus. nat. Hist.*, *106*, 173–220. (Occurrence and attributes of pentagonal symmetry)

57 Woodland, W., 1906–7. *Quart. J. micr. Sci.*, *49*, 305–25, 533–59, and *51*, 45–54, 483–509. (Spicule formation)

SPINES AND PEDICELLARIAE

58 Fujiwara, T., 1935. *Annot. Zool. Japon*, *15*, 62–9. (Poisonous pedicellariae of *Toxopneustes*)

59 Pérès, J. M., 1950. *Arch. Zool. exp. gén.*, *84*, *Notes et revue*, *3*, 118–36. (Glandular pedicellariae of *Sphaerechinus*)

60 Perrier, M. E., 1870. *Ann. Sci. nat.*, *13*, 5–81. (Pedicellariae of asteroids and echinoids)

61 Romanes, G., and Ewart, J., 1881. *Phil. Trans.*, B, *172*, 829–85. (Pedicellariae of asteroids and echinoids)

62 Uexküll, J. von, 1899. *Z.Biol.*, *37*, 334–403. (Physiology of pedicellariae)

THE TUBE-FEET

63 Kerkut, G. A., 1953. *J. exp. Biol.*, *30*, 575–83. (Forces exerted by tube-feet)

64 Kerkut, G. A., 1954. *Behaviour*, *6*, 206–32. (Coordination of tube-feet)

65 Kerkut, G. A., 1955. *Behaviour*, *8*, 112–29. (Retraction of tube-feet)

66 Mortensen, T., 1937. *Biol. Medd., Kbl.*, *13*, 1–28. (Fossil disk rosette from Jurassic)

67 Nichols, D., 1959*a*. *Quart. J. micr. Sci.*, *100*, 73–88. (Tube-feet of *Echinocardium*)

68 Nichols, D., 1959*b*. *Quart. J. micr. Sci.*, *100*, 539–55. (Tube-feet of *Echinocyamus*)

69 Nichols., D., 1960. *Quart. J. micr. Sci.*, *101*, 105–17. (Tube-feet of *Antedon*)

70 Nichols, D., 1961. *Quart. J. micr. Sci.*, *102*, 157–80. (Tube-feet of *Cidaris* and *Echinus*)

71 Paine, V. L., 1926. *J. exp. Zool.*, *45*, 361–6. (Adhesion of tube-feet)

72 Paine, V. L., 1929. *Amer. Nat.*, *63*, 517–29. (Tube-feet as autonomous organs)

73 Réaumur, M. de, 1712. *Hist. Acad. roy. Sci.*, *4*, 115–45. (First account of tube-feet)

74 Reichensperger, A., 1905. *Z. wiss. Zool.*, *80*, 22–55. (Tube-feet of stalked crinoids)

75 Smith, J. E., 1937. *J. Mar. biol. Ass. U.K.*, *22*, 345–57. (Structure and function of various tube–feet)

76 Smith, J. E., 1946. *Phil. Trans.*, B, *232*, 279–310. (Starfish tube-foot/ampulla system)

77 Smith, J. E., 1947. *Quart. J. micr. Sci.*, *88*, 1–14. (Action of asteroid suckered tube-feet)

LARVAL FORMS

78 Bury, H., 1889. *Quart. J. micr. Sci.*, *29*, 409–49. (Embryology of echinoderms)

79 Chadwick, H. C., 1914. *L.M.B.C. Memoir*, *22*. (Echinoderm larvae)

80 Dawydoff, C., 1948, in *Traité de zoologie*, ed. P.-P. Grassé. (Development and larval forms)

81 Fell, H. B., 1945. *Trans. roy. Soc. N.Z.*, *75*, 73–101. (Embryology)

82 MacBride, E. W., 1914. *Textbook of embryology.* Vol. I *Invertebrata*

83 Mortensen, T., 1931, 1937, 1938. *Mém. Acad. R. Belg.*, ser. 9, *4*, 7. (Studies on larval forms)

84 Nyholm, K.-G., 1951. *Zool. Bidr. Uppsala*, *29*, 239–76. (Development of *Labidoplax*)

85 Rees, C. B., 1953. *J. Mar. biol. Ass. U.K.*, *32*, 477–90. (Larvae of spatangoids)

86 Seelinger, O., 1892. *Zool. Jb. Anat.*, *6*, 161–444 (Development of *Antedon*)

87 *Thorson, G., 1950. *Biol. Rev.*, *25*, 1–45. (Larval ecology)

For a prolonged, and in places heated, correspondence on the phyletic importance of echinoderm larval forms between F. A. Bather, W. Gemmill, E. W. MacBride and T. Mortensen from 1920 to 1923, see *Nature, Lond.*, *107*, 132–3; *108*, 459–60; *108*, 529–30; *108*, 530; *110*, 806–7; *111*, 47; *111*, 47–8; *111*, 322–3; *111*, 323–4; *111*, 397.

CYSTOIDEA AND BLASTOIDEA

88 Barrande, J., 1887–99. *Système silurien du centre de la Bohême* (*I*), 7, 1–233. Ordre des Cystidées. Prague.

89 Bather, F. A., 1913. *Trans. roy. Soc. Edinb.*, *49*, 359–529. (Caradocian cystoids from Girvan)

90 Chauvel, J., 1941. *Mém. Soc. géol. Bretagne*, *5*, 1–286. (Review of thecal canal systems in cystoids)

91 Croneis, C., and Geis, H. L., 1940. *J. Pal.*, *14*, 345–55. (Ontogeny of the blastoids)

92 Etheridge, R., and Carpenter, P. H., 1886. *Catalogue of the Blastoidea.* British Museum (Natural History). London.

93 Jaekel, O., 1889. *Stammesgeschichte der Pelmatozoen.* 1, Thecoidea und Cystoidea. (*Echinosphaerites*)

94 Jaekel, O., 1918. *Palaeont. Z.*, *3*, 1–128. (Structure of diplopores)

95 Regnéll, G., 1945. *Medd. Lunds geol.-min. Instn.*, *108.* (Modern classification of cystoids)

96 Spencer, W. K., 1938, in *Evolution*, essays to Goodrich, ed. G. R. de Beer. Oxford. (Structure and mode of life of of some cystoids)

EDRIOASTEROIDEA AND HETEROSTELEA

97 *Bassler, R. S., 1935. *Smithson. misc. Coll.*, *93*, no. 8, 1–11. (Classification of edrioasteroids)

98 Bather, F. A., 1915. *Geol. Mag.*, *VI*, *2*, 211–15, 259–66. (Morphology and bionomics of edrioasteroids)

99 Bather, F. A., 1925. *Palaeont. Z.*, *7*, 1–15. (*Cothurnocystis*)

100 Bather, F. A., 1930. *Arch. ital. Zool.*, *14*, 431–9. (Mode of life of heterosteles)

101 Gislén, T., 1930. *Zool. Bidr. Uppsala*, *12*, 199–304. (Affinities between heterosteles and chordates)

102 Jaekel, O., 1921. *Palaeont*, *Z.*, *3*, 1–128. (Multiple gonopores of *Cothurnocystis*)

103 Spencer, W. K., 1938, in *Evolution*, essays to Goodrich, ed. G. R. de Beer. Oxford. (Reconstruction of *Gyrocystis*)

ENIGMATIC GROUPS

104 Bather, F. A., 1926. Introduction to Withers, 1926. (Machaeridia)

105 Bather, F. A., 1929. 'Echinoderma' in *Encycl. Britt.*, *14th ed.*, *7*, 894–904. (Machaeridia)

106 Fedotov, D. M., 1926. *Proc. zool. Soc. Lond.*, *1926*, 1147–57. (Structure and systematic status of ophiocistioids)

107 Hecker, R. F., 1938. *C. R. (Doklady) Acad. Sci. U.R.S.S.*, *19*, 425–7. (The ophiocistioid *Volchovia*)

108 Regnéll, G., 1948. *Norsk Geol. Tidsskr.*, *27*, 14–58. (Ophiocistioids)

109 Sollas, W. J., 1899. *Quart. J. geol. Soc. Lond.*, *55*, 692–715. (Ophiocistioids)

110 Sollas, I. B. J., and Sollas, W. J., 1912. *Phil. Trans*, B, *202*, 212–32. (Ophiocistioids a class)

111 Ubaghs, G., 1953, in *Traité de paléontologie*, *3*, ed. J. Piveteau. (Ophiocistioids)

112 Whitehouse, F. W., 1941. *Mem. Queensland Mus.*, *12*, 1–28. (Cycloids and cyamoids)

113 Withers, T. H., 1926. *Catalogue of the Machaeridia*, British Museum (Natural History). London.

INVERTEBRATE RELATIONS OF ECHINODERMS

114 *Baldwin, E., and Yudkin, W. H., 1949. *Proc. roy. Soc.*, B, *136*, 614–31. (Phosphagens of invertebrates and protochordates)

115 *Bermann, W., 1949. *J. mar. Res.*, *8*, 137–76. (Sterols of marine invertebrates)

116 Bury, H., 1895. *Quart. J. micr. Sci. 38*, 45–136 (Pentactula theory)

117 Clark, A. H., 1922. *Smithson. misc. Coll.*, *72*, *11*, 1–20. (Echinoderms as aberrant arthropods, see also review by Bather, F. A., 1922, *Nature, Lond.*, *109*, 640–1)

118 Grobben, K., 1923. *S. B. Akad. Wiss. Wien*, *132*, 263–90 (Echinoderm-hemichordate relations)

119 *Hyman, L. H., 1959. *The invertebrates*, vol. V. *Smaller coelomate groups*. New York. (Deuterostome relationships)

120 Jensen, D. D., 1960. *Nature, Lond.*, *188*, 649–50. (Supposed relationships of nemertines and chordates)

121 *Kerkut, G. A., 1961. *Implications of evolution*. London. (Invertebrate relationships)

122 Marcus, Ernesto, 1958. *Quart. Rev. Biol.*, *33*, 24–58. (Evolution of the animal phyla)

123 Semon, R., 1888. *Jena. Z. naturw.*, *22*, 175–309. (Dipleurula and pentactula theories)

ECHINODERMS AND THE ORIGIN OF CHORDATES

124 Berrill, N. J., 1955. *The origin of vertebrates.* Oxford.

125 Bone, Q., 1960. *J. Linn. Soc.* (*Zool.*), *44*, 252–69. (Origin of chordates from hemichordates)

126 Carter, G. S., 1957. *Syst. Zool. 6*, 187–92. (Not paedomorphosis)

127 Fell, H. B., 1940. *Nature, Lond.*, *145*, 906–7. (Origin of vertebrate coelom)

128 Fell, H. B., 1948. *Biol. Rev.*, *23*, 81–107. (Echinoderm embryology and the origin of chordates)

129 Garstang, W., 1894. *Zool. Anz.*, *17*, 122–5. (Echinoderms to chordates by paedomorphosis)

130 Garstang, W., 1928. *Quart. J. micr. Sci.*, *72*, 51–187. (Position of tunicates in echinoderm—to—chordate line)

131 Gislén, T., 1930. *Zool. Bidr. Uppsala*, *12*, 199–304. (Affinities between heterosteles and chordates)

INDEX

Page-numbers in **heavy type** refer to illustrations

Page-numbers in *italic* refer to the definition of a term, and/or the chief reference to it